CRITICAL PLANETARY ROMANTICISM

CRITICAL PLANETARY ROMANTICISM

RELIGIOUS AND SCIENTIFIC SOURCES FOR A NEW MATERIALISM

WHITNEY A. BAUMAN

Columbia University Press *New York*

Columbia University Press
Publishers Since 1893
New York Chichester, West Sussex
cup.columbia.edu

Cataloging-in-Publication Data is available from the Library of Congress.

ISBN 9780231211826 (hardback)
ISBN 9780231211833 (trade paperback)
ISBN 9780231559089 (epub)
ISBN 9780231566049 (PDF)

LCCN 2026025916

Printed and bound by CPI Group (UK) Ltd, Croydon, CR0 4YY

Cover design: Noah Arlow
Cover image: Ernst Haeckel

GPSR Authorized Representative: Easy Access System Europe,
Mustamäe tee 50, 10621 Tallinn, Estonia,
gpsr.requests@easproject.com

CONTENTS

III RETHINKING GOODNESS (PERSPECTIVALISM)

ILLUSTRATIONS

ACKNOWLEDGMENTS

I started thinking about this book around 2009, when I learned that Ernst Haeckel coined the term *ecology* and that he was also an accomplished artist and armchair philosopher. This recognition was thanks to listening to a paper by Evan Berry at a conference I hosted at Florida International University (FIU) through a cooperative grant from the American Academy of Religion. Even though the ideas for this book were emerging at that time, I didn't really turn my attention to it until I received a sabbatical from FIU and a Humboldt Fellowship from 2015 to 2017 to do research in the Ernst Haeckel archives at the Haeckel Haus (a.k.a. the "Villa Medusa"), Schiller Universität, Jena, Germany. I'd especially like to thank the late Olaf Breidbach for supporting my Humboldt application and Thomas Bach for being a gracious host and for his help at the Haeckel Haus. Those two years were invaluable in terms of research, language skills, and mapping a huge chunk of what I thought would be this book. Little did I know the circuitous route this work would take. What started as a book about the religiosity at the heart of modern ecology, especially through the work of Ernst Haeckel, has turned into a book about the connections between "Romanticisms" of the nineteenth and twentieth centuries, mainly in Europe, and the so-called new materialisms. Though Haeckel still plays an important role in the texts, and the critiques of modern, reductive and productive materialism still remain, the book took on its own life and I just kind of followed. There are several people and organizations to thank for the final product (the shortcomings of which are, of course, all mine).

First, chapter 4 is a revised version of a chapter that appeared in Sigurd Berman and (the late) Forest Clingerman's book, *Arts, Religion, and the Environment* (Brill, 2018). Second, the work on Rachel Carson in later chapters is due in large part to grants received from the Wolfsonian Humanities Public Lab to do research in the Carson archives at Yale. Some of that work was also done in cooperation with my colleague Amanda Nichols ("The Watery Depths of American Environmentalism: Marjory Stoneman Douglas, Rachel Carson, and Sylvia Earle," *Journal for the Study of Religion, Nature and Culture* 16, no. 3 (2022): 394–425); and Kevin O'Brien (*Environmental Ethics and Uncertainty: Wrestling with Wicked Problems*, Routledge, 2019). I thank them both for these invaluable collaborations. Third, some of the thinking of this book was also developed in chapters that appeared in two additional books: Peter Scott, Kate Rigby, and Sigurd Bergmann, *Religion, Materialism and Ecology* (Routledge, 2023); and Jerome Stone and Donald Crosby, *The Routledge Handbook of Religious Naturalism* (Routlege, 2018).

In addition to these more direct connections to this text, I am always supported by the work of the Forum and Religion and Ecology, and especially the work of Mary Evelyn Tucker and John Grim. My thinking is also always stimulated by my colleagues and friends at the International Society for the Study of Religion, Nature and Culture . Finally, my work with the Religion and Ecology unit at the American Academy of Religion

(AAR) and with the Religion and New Materialisms Seminar of the AAR have been invaluable. I especially want to thank my co-collaborators in the leadership of that seminar: Karen Bray, Heather Eaton, Mary Evelyn Tucker, Catherine Keller and Jeorg Rieger. The discussions in that seminar and with many individuals both who are listed here and some who are not have been invaluable in producing this text. Some of the thinking in this book can also be found in the book that emerged from that seminar, *Earthly Things* (Fordham University Press, 2023).

Along with these major organizational and collegial influences, a number of other conferences and colleagues have helped nurture this book to fruition. Donovan Schaefer and a conference at the Ian Ramsey Center, multiple conferences of the Institute on Religion in an Age of Science, a conference hosted by Lisa Sideris on wonder at Indiana University, Bloomington, an International Whitehead conference in the Azores, a conference hosted by Matthew Wickman at Brigham Young University, a seminar at Humboldt University hosted by Devin Zuber (who also gave me serious feedback on earlier drafts of this book), a series of conferences hosted by James Miller at Duke Kunshan University in China (who has also been a great co-collaborator), a symposium on "dark green religion" in Mainz, Germany organized by Bron Taylor, and seminars and courses at both the University of Gadjah Mada, in Jogjakarta, Indonesia, and at Florida International University in Miami, Florida. All these interlocutors can be implicitly found in this pages of this work.

I would especially like to thank two major long-term collaborators and friends. One is my colleague and friend Lisa Stenmark, with whom I have worked on a number of projects including coediting a series at Bloomsbury (formerly Lexington Books), Science and Religion as Critical Discourse. Our friendship and collaboration of over twenty years has been invaluable. The second is my cofounder, co-collaborator, and friend at *Counterpoint: Navigating Knowledge* in Berlin, Kocku von Stuckrad. Our friendship and collaborations over the last decade have been invaluable.

I would be remiss if I didn't also thank the reviewers of earlier drafts of this manuscript, some of whom I know (Catherine Keller and Kate Rigby), and others whom I do not know (anonymous). And, of course, this book would be impossible without the attention and encouragement from Wendy Lochner at Columbia University Press. The constructive criticisms and feedback in this process have made this a much better text.

Finally, since this book was started, so much has happened: deaths, pandemics, the rise of nationalisms, increasing intensity of climate change, and so many other things. I want to honor my ancestors, in particular my late father, Lynn Bauman, and late uncle Phil Roberts. I also want to acknowledge and honor those close friends who passed away during the writing of this book: Terry Zamojski, Harmony Abrams, Sean McAuley, and Gaymon Bennett. Finally, I'd like to thank my late dissertation advisor, Rosemary Radford Ruether, who was an example of thinking planetarily if there ever was one. I know many readers have experienced trauma over the past couple of years, so I just want to name the trauma that has gone into the production of this book: the political trauma of rising rightwing nationalisms as well as the ecological trauma/anxiety/despair of increased devastation from the effects of overconsumption, overextraction, and climate change.

My friends, family, students, and colleagues around the world give me hope that we can have a different planetary future. They remind me that we can find little island shelters of hope in what seems like an overwhelming destructive hurricane, and I thank them greatly for that. This book is written in the space of those little islands of hope that we can, as long as we are still breathing, carve out different planetary paths toward the future. One such proponent of that hope, Jane Goodall, died right before I submitted the final version of this text. I dedicate this book to her and the planetary vision that she fought for throughout her life.

CRITICAL PLANETARY ROMANTICISM

INTRODUCTION

Rethinking Nature, Religiously—Reattuning to Our Worlds

There are still many philosophers who would construct the world out of their own inner consciousness, and who reject our empirical science precisely because they have no knowledge of the real world. On the other hand, there are many scientists who still contend that the sole object of science is 'the knowledge of facts, the objective investigation of isolated phenomena'; that 'the age of philosophy' is past, and science has taken its place.

. . . Monism, on the contrary (likewise taken in its widest sense), recognizes one sole substance in the universe, which is at once 'God and nature'; body and spirit (or matter and energy) it holds to be inseparable. The extra-mundane God of dualism leads necessarily to theism; and the intra-mundane God of the monist leads to pantheism.

—Ernst Haeckel, *The Riddle of the Universe at the Close of the Nineteenth Century*

It is standard in the world of religious studies (and the humanities in general) to study the history of interpretation: to recognize that ideas and thoughts cannot be cut and pasted from the past into the present without historical, sociological, and other forms of critical analyses. The context in which the writer writes is extremely important for understanding any text. The need for the study of the history of interpretation in the natural/physical sciences, however, is not so widely assumed (at least outside of the academic humanities). Science is often quoted as if it is unchanging or as if it somehow was and is always in agreement with the most current version of science (a form of presentist thinking—the idea that *now* we, Western moderns, (WMs) have the most correct knowledge that we WMs have ever had). Copernicus, Newton, Einstein, and Hawking are selectively quoted as if they are speaking about the same thing, as if their knowledge moved from older to newer in a progressive manner, and as if scientific knowledge comes from a place of objectivity against the relativity of religion and culture. Of course, to successfully do this, a certain version of Copernicus, Newton, Einstein, and Hawking must be quoted out of context. The socio-historical context of the actual scientists is left behind in favor of the elements of "scientific narrative" that fits with whatever science is current. This creates a naturalistic, scientific worldview that seems to extend from (at least) the so-called Scientific Revolution to the present in a progressive narrative. Yes, there are shifts and changes, but the building blocks are all there and the context of the scientists and mistakes of the past are overlooked for the important role that past "discoveries" natural philosophers-scientists made in support of the most present understanding of the natural world and the universe.

This way of building narratives falls prey to what I refer to as the reductive and productive model of progress. Within such a narrative, one can read progress by reducing data (historical and contemporary) to the necessary bits that help one continue on in a single narrative of progressive time, while "backgrounding" the data that do not fit.[1] Such progressive time in our era of globalization and climate change is fossil-fueled. In a sped-up, fossil-fueled time, the most efficient knowledge, the knowledge that doesn't slow down the pace of progress, is that which is reductive and productive. However, I would argue this reductive productive model of knowledge production in the "sciences" broadly conceived ("human" and "natural") has not been a dominant model even in the so-called modern, Western traditions.[2] Human understandings of knowledge and knowledge production have largely to do with human understandings of our relationship to the rest of the natural world. I argue that models of reductive and productive knowledge depend upon transcendent understandings of knowledge and ideas of the really real (metaphysics and ontology), and that more immanent understandings of knowledge, based upon a type of nonreductive naturalism, produce knowledge that leads to awe, wonder, and concern (aesthetics and ethics) about the world, rather than merely knowledge of what things in the world can do for us. Contemporary "new materialisms" and "neo-animisms" (although it is not clear what is "new" or "neo-" about them other than WMs are engaging with these older traditions) are articulating this type of nonreductive naturalism, but there are also historical roots for thinking within such immanent frames, at which we ought also to look.

I open this book with two quotes from Ernst Haeckel because he coined *ecology* as a science, and he was not a reductive, productive scientist. Rather he, like many scientists of his era, was Romantic and wholistic, or what Dalia Nassar calls a "romantic empiricist."[3] The "origins of ecology" as a science (though clearly not reducible to the work of Haeckel alone) provide us with a nonreductive model of science, and perhaps even with a form of "religious naturalism," or "eco-spirituality." In what follows, I argue for a troubling of progressive, reductive, and productive readings of knowledge and reality, and through the lens of earlier Romantic scientists such as Ernst Haeckel suggest that interpreting nature, including the work of science, is every bit as hermeneutical as interpreting any other "text." Reinterpreting ideas from religion and science is necessary for reattuning to the evolving worlds in which we live. In this book, I look at some Romantic scientists as early developers of a non-reductive "religious naturalism," a form of "new materialism" or what I will call a critical planetary romanticism (CPR) for the Earth. Haeckel's (and other's) evolutionary and ecological views of the world force those of us who find them compelling to reattune to the worlds in which we live in ways that make humans and all things human a part of the rest of the evolving planet in which we live. Of course, this reattuning is never done outside of historical, political, socioeconomic, gendered, sexed, and racial contexts. All these factors help to shape science in the given time and place in which it is done. These things are, in other words, always entangled together into interpretations of the worlds in which we live.[4]

Haeckel and other Romantic scientists of his era were engaged in articulating an understanding of the type of worlds the (then) emerging natural sciences might describe. For some of these Romantic scientists or "Romantic empiricists" the early naturalistic worldview was neither a reductive naturalism nor a mystical wholism, but a form of Romanticism. Haeckel and other Romantic scientists were constructing this naturalistic worldview during a time in which the epistemological authority (and political power) of theology was being challenged across Europe, including Britain, due to the ongoing effects of scientific revolutions and Protestant Reformations, the era of

European colonization, and the Industrial Revolution. Indeed, religious leaders were no exception to the effects of these changing intellectual authorities. Herder had already begun an "eco-hermeneutics" inspired by his revised version of Spinoza.[5] The so-called clerical naturalists were already spreading nature practices based upon natural theology, especially the idea that close study of creation was a way to know more about God. Their work at local levels would eventually help spread ideas emerging out what would become the natural sciences.[6]

An evolutionary perspective of the origins of life and of human beings was just coming into focus with contributions from the German and British Romantics (particularly Goethe) and from scientists such as Lamarck, Wallace, Erasmus Darwin and (of course) Charles Darwin. At this time, there were no clear methodological or epistemological grounds that brought together the various observations from medical, anthropological, biological, geological, chemical, and physical sciences into a single naturalistic worldview, much less one that united the "natural" and "human" sciences.

There were several different viable models competing to ground the emerging *Naturwissenschaftliche Weltanschauung* (naturalistic worldview), ranging from dualism, idealism, and materialism to Haeckel's preferred form of "triune monism." And, within each of these main options, there were several different subcategories. This book excavates the detritus of some of these nonreductive and nondualistic naturalisms and brings them into conversation with emerging contemporary critical discourses on nonreductive forms of naturalism, new materialisms, and critical theories, in order to develop what I am calling a CPR for the earth. A CPR might help break us out of the human exceptionalism found in modernity, and to a great extent in postmodernity and colonial and postcolonial discourses, both in the human and natural sciences. It might help us break through the progressive understanding of time and the reductive and productive model of knowledge that seems to be tied to it, which when read into nature has underscored racists and heteropatriarchal understandings of the human. These critical "post-" forms of thought, even though providing much critical analysis of "the human," have until recently mostly ignored the more than human world and our deep embeddedness in that world. Noted exceptions being the work of such scholars as Donna Haraway, David Wood, Freya Matthews, thinkers on process thought and deep ecology, and a few others. A CPR recognizes that once we are placed in an evolutionary and ecological context all our thinking (scientific or otherwise) must also involve hermeneutics: we cannot get an objective view of bare facts because we are emergent in the ongoing process of nature naturing. CPR must also include an encounter with the ways in which other types of embodiments, life-forms, and their world-frameworks experience the world around them. The only thing that makes up the whole is the continual manifestation of its interacting parts. As Freya Matthews argues about her version of panpsychism:

> In this sense panpsychism entails a pervasive spirituality: the One is a party to all our undertakings, from the most large-scale to the most seemingly trivial, and the meaning of our actions and our lives is never fixed, but is continually being negotiated and renegotiated through our ongoing conversations with the Many.[7]

Critical theories should be thought of in the context of this wider planetary community rather than limited to human experiences.

A second argument of this book is that it is time to seriously engage (again) multiple models for understanding the natural world. My argument, in sum, is that the reductive materialistic and production-based model of nature is not the assumed "norm" of modern science; rather, it is a twenty-year blip between World War II and the early 1960s. Indeed this has never been the norm for many practicing scientists themselves, who are often inspired by the awe and wonder produced by the study of nature; it is, rather, I will argue a product of the increased speed at which neoliberal capitalism is globalized during World War II and after, a period some are referring to as the "great acceleration,"[8] which is, of course, acceleration for some and deceleration for others.[9] The globalization of the reductive productive model is selected as the norm by neoliberal capitalism and neoliberal capitalists precisely because it is an economically efficient model, and it lends itself to a progressive understanding of history for some (even in the face of a more "relational" understanding of nature coming from sciences such as ecology, evolutionary theory, cosmology, and quantum physics). As Mel Chen notes, this efficiency, speed, and progress creates and projects slowness onto other bodies.[10]

In this text when I refer to the *reductive and productive models*,[11] I am mostly arguing here about the biological, physical, and earth sciences, and, by extension, how other sciences are taken up in the project of developing a reductive naturalistic worldview. Physics, for instance, though never fully and only reductively, was (via Newton) influenced by a certain theological dualism and mechanism much earlier than biological sciences. The theological dualism between body and spirit held by some theologians morphed into the matter and energy dualism in general. This dualism also carried with it the idea of matter as passive and energy/spirit as active. Despite the efforts of some reductive scientists to fit the life sciences into this physics-based model, it never fully took hold in a major way, or so I am arguing, until the period from World War II up until the mid-1960s. By this time Einstein's revolution in physics, the emerging environmental crisis outlined in Rachel Carson's *Silent Spring*, and the increasing cultural and economic globalization that spread pluralistic understandings of religions and cultures around the world (though not equally), had all begun to unravel the reductive, productive, and progressive model. There are, perhaps, "better" models, as in less ecologically and socially destructive ways of thinking about humans and nature than the reductive-productive-progressive model, and this book hopes to contribute to those alternatives. Rethinking these models requires reattuning to the worlds in which we live and taking account of the violence caused by the reductive-productive-progressive model of worlding.

By way of introduction, I think it will be most helpful to start from where we "are" and move backward to the nineteenth-century German Romantic, context. After all, we all think and understand the world through our own historically and ecologically embodied location, so it is better to start with that context before sailing off into waters less familiar. I focus on the German Romantics (scientists, theologians, and philosophers) because many of them were seeking to find a way to integrate ideas and culture back into a single evolving plane of life. In this way, I do think they are important predecessors for these (so-called) newer forms of materialisms that attempt the same. Furthermore, we might learn from the pitfalls of these earlier Romantics through engaging with their works and how they were taken up into ideas, politics, and policies in problematic ways.

REFLECTIONS ON THE (SO-CALLED) ANTHROPOCENE NARRATIVE

Being cannot be anything but being-with-one another, circulating in the with and as the with of this singularly plural coexistence.

—Jean-Luc Nancy, *Being Singular Plural*

The last seventy years have been the most industrialized the planet has ever seen for many places, and with that industrialization has come progress for some people but also problems for many more (human and more than human). The most common catchall name for the problems that have resulted from the industrialization of the planet, or better yet, the fossil-fueled speed of the planet is, of course, global climate change.[12] I refer to this phenomena, following Hunter Lovins, as "climate weirding" rather than climate change.[13] The basic reason for this is that change can be interpreted as merely constant and "normal"; it is just what happens. Weirding, though not necessarily good or bad, provides us with a moment of pause, a chance to take note of change in a way that is not just routine: weird things help us to pause from our normal experience of time and reflect on our relationships to phenomena.

With climate weirding, global changes will not be everywhere equal in the same direction; they will be different and varied. A hot summer here, a warm or very cold winter there, an extremely long drought here, a large hurricane there, a global pandemic from time to time, all dotted with what we think of as "normal" weather and life in between. The problem with our climate is that the patterns of seasons and weather we know, and which have been relatively stable over the last ten thousand years in which human civilizations have emerged and thrived, are becoming chaotic. They are getting weird. The climate is, as Timothy Morton notes, a hyperobject: it is of a scale that the human mind and experience cannot see or comprehend all together, but rather we only see and feel glimpses of its effects.[14] This has made it harder to deal with the effects of the "slow violence" caused by climate weirding.[15] We simply don't experience the climate all at once in our daily lives in which we do experience change on the regular. Mere "change" does not seem to capture the weirdness of these effects on our daily lives.

Recently scientists, historians, and environmental activists, as well as NGOs and government organizations, have begun to rally around the idea that we are in the new geological era of the Anthropocene, even if this term was rejected by the International Union of Geological Sciences in 2024. For reasons that overlap with my objections to the term *climate change*, I also dislike the term and concept of the Anthropocene. First, it papers over the vast differences in human responsibility for climate weirding, by lumping the Anthropos all together. In a way this re-creates the false, liberal Lockean ideal that we are born with a tabula rasa mind and have equal opportunities to "make it" in the world. Given the gross economic inequities throughout the world, the Black Lives Matter and #MeToo movements, decolonial movements, and the Indigenous rights movements that seek justice for past and current wrongs and inequalities, the idea that we can be lumped together as equally responsible for climate change and environmental ills under "the Anthropocene" is a gross oversight at best and an insult at worst. Not all peoples are equally responsible for pumping fossil fuels and other greenhouse gasses into the air, and many places that are "progressing" toward modernity are now being told by well-meaning environmentally minded people they can't or shouldn't industrialize:

this is, of course, a form of "green colonialism." We live in and co-create many different worlds, not just one. For this reason, I and others prefer calling our era the *fossil-fueled era* or something to that effect.[16] This type of nomenclature at least hints at the problems we face is because of the speed and pace at which fossil-fueled worlds proceed, and the destruction they leave behind. Again, speed for some and stasis for others: not all participate in this fossil-fueled reality equally, but all humans (and planetary creatures and systems) are affected by it.

At the same time, the idea of "climate change" and the *Anthropocene* as the new name for the current geological era join together to suggest an amount of human agency and responsibility over the world that I am not sure is warranted or even credible.[17] Can't humans just change the climate back with technology? Isn't the "human era" announcing the arrival of an anthropocentrism come to fruition? Donna Haraway argues for this era to be called the *Chthulucene*.[18] She thinks the Anthropocene reasserts this idea of humans as somehow isolated individuals, as separate and distinct from the rest of the natural world, and that we should instead think of the "monstrous" hybrid ways we become with other planetary creatures. She writes, "To be animal is to become-with bacteria (and, no doubt, viruses and many other sorts of critters; a basic aspect of sympoiesis is its expandable set of players)."[19] We need to complicate the narratives of the Anthropos, in other words.

Strangely enough, there is a narrative that feeds into this Anthropocene idea that is also at the heart of some modern Western environmental narratives about our current situation: viz. the story told of the Scientific Revolution and the death of nature.[20] I am deeply in debt to Carolyn Merchant's groundbreaking work on the "death of nature," and she has played a huge part in the development of my own and many others' thought around humans-and-the-rest-of-nature; yet, even this environmental narrative has become a source for making humans and human agency exceptional to the rest of the natural world.[21]

FROM THE DEATH OF NATURE TO THE ANTHROPOCENE: BENDING OUR WORLDS TOWARD THE ANTHROPOS

Carolyn Merchant's argument in *The Death of Nature* has been a strong one for environmental philosophers and historians, historians of science, and feminist environmentalists. Many different scholars have argued along these similar lines. The basic arc of the narrative is that there was once a time (before the modern Western world) when humans and nature were more interrelated and nature was even "enchanted," and then with the modern Scientific Revolution and European colonization came the long process of the mechanization of nature, which turned matter into dead stuff. Her later works, including *Reinventing Eden* and *Autonomous Nature*, are critical of some of the environmental and other narratives of decline, as well as "recovery" and "progress." In an effort to help us break out of the confines of such narratives she suggests forming "partnerships" with the rest of the natural world: beyond linear and singular narratives.[22]

The problem with singular narratives is that a single story is incapable of capturing the chaotic and complex worlds in which we live. Again premodern/precolonial, modern/colonial, and postmodern/postcolonial nomenclature still carries on this type of singular narrative of time. However, counternarratives are the norm if we look at global human histories. We ignore the rich diversity of narratives to the peril of our imaginative-embodied realities.[23] There is a way in which the "death of nature" narrative leads seamlessly into "the Anthropocene" narrative.[24] Though not Merchant's intention, there is a way in which the arrival of the Anthropocene can confirm the "death of nature" in an almost fatalistic and positive way. The idea of history moving from the death of nature to the Anthropocene repeats the colonial logic that Walter Mignolo identifies in *The Darker Side of Western Modernity*: all histories and cultures are arranged around a single "tunnel of time."[25] This, I argue (along with other post-/decolonial thinkers), creates a dangerous logic of domination and colonial violence. Furthermore, it is just the latest version of the "anthropic principle," which reorders our understanding of geohistory (which includes the history of the earth and human cultures all together) and even universal history to lead up to the present moment, as if it were inevitable.

Even among those ultimately concerned with contemporary ecological crises, the Anthropocene seems to confirm that the problem and solution is found in one place, the Anthropos, thus recapitulating a form of eco-colonization.[26] Again, Carolyn Merchant, in her book *Autonomous Nature*, understands this problem fully and indeed has been arguing for years that the sciences of chaos and complexity demand that we understand the entire planetary community as living and agential in nature. Most of her writing since *The Death of Nature* has been an effort to help us reattune to the worlds described by chaos and complexity theories. Such a nature cannot be controlled like a deviant person, or saved like a helpless child; rather, we must enter into partnerships (sometimes messy and abusive ones) with the rest of the natural world. Indeed, as Bruno Latour notes, what we need is the death of the concept of nature (and culture) and to start thinking about ongoing creation instead.[27] We need, according to him, to be better materialists. "We were supposed to have been materialists during the modern period! In fact, we were no such thing; we were dematerializing ourselves, abstracting ourselves, and conceptualizing an abstract world—one that has a lot of useful functions within scientific networks—but that is not the earth. With the new ecological situation, we are coming back down to earth."[28]

This current book, then, can also be seen as another attempt to trouble the narrative that moves from the death of nature to the Anthropocene, and to continue troubling progressive, declensionist, and recovery narratives in favor of timelines that suggest no single teleology but rather follow the logic of something like starling murmurations: If one follows the murmuration of starlings, there are clear and definite patterns that form, but these are sympoietic; they are not predetermined, but rather emergent in the process of individual birds acting in concert with the others that make up the whole, or not.[29] There is no definite direction in which the murmuration might go, but rather it meanders.[30] Such a metaphor, I will argue throughout this book, is a better way to understand time, nature, and human relationships as a part of nature, than those metaphors that suggest some sort of directionality. A CPR of the type I develop in this book will allow for multiple narratives to flourish in ways that understand we live in worlds among worlds, all of which at any given moment make up the evolving planetary community.[31] There is no single formed whole, but rather multiple worlds. As mentioned above regarding Merchant's work, it is important to look at the many alternative narratives, methods, and ideas one finds. For the purposes of this book, one such tradition that has challenged teleology, single interpretations, productive and reductive models of science, progressive understandings of time, and human exceptionalism is that of the Romantics. Scientists influenced by German Romanticism, though at times very problematic, provided earlier models that challenged reductive models of nature, which had not yet fully taken hold of the WM's mindset.

TRIPPING UP ANTHROPIC NARRATIVES: REATTUNING TO EARTHLY VOICES

Ernst Haeckel held one such nonreductive vision of nature in his idea of triune monism. Like Romantic science in general, this concept was developed from within the so-called *Kulturkampf* (cultural battles) of the nineteenth century between theologically informed foundations for knowledge and politics, and other foundations for knowledge and politics emerging from what we would now call the natural sciences. Haeckel began his own life from within a liberal, Protestant family influenced by the theological ideas of Friedrich Schleiermacher and other liberal theologians who sought to demystify Christianity and the Bible.[32] Schleiermacher, drawing from Herder's work,[33] attempted to describe Christianity from an existential perspective and argued that the fundamental condition described by the Christian story of creation and salvation was one of dependency upon God. We are, so to speak, thrown into a world that is not of our own creation and dependent through our entire lives upon others (and ultimately the great Other) for our existence (a form of entanglement to be sure).[34] Haeckel digested this insight, but for him God would be switched out (as it was for Spinoza and many others before him) for Nature: humans, in other words, are "thrown" into nature, and dependent upon nature and the processes of nature. Nature, for Haeckel, then is Schleiermacher's God crossed out.

Haeckel was also enamored by the first wave of "historical Jesus" scholarship in the nineteenth century that sought to understand the life of Jesus as a historical figure (rather than as some sort of divine, supernatural figure).[35] Ludwig Feuerbach's more psychological understanding of theology as collective human projection was also quite important for Haeckel. Feuerbach argued that whatever else "god" or "ultimate concern" might be, it is primarily a collective human projection of narrative and order onto the rest of the natural world and therefore not about metaphysics and ontology but psychology, sociology, history, and ethics.[36] All theological understandings of God are then about theological anthropology, or human self-understanding. This move toward theological anthropology helped Haeckel place philosophy and theology within his naturalistic worldview as part of "science." At this time in Germany, one must remember that the "sciences" included both natural sciences and the sciences of the spirit, what today we might call the humanities. The German words *Geisteswissenschaften* and *Naturwissenschaften* understand science in its broadest sense to include all knowledge production.[37]

Finally, Haeckel's scientific understanding of the natural world was influenced heavily by a line of immanent, pantheistic thinking stretching from Bruno to Spinoza, to Alexander von Humboldt, Goethe, Herder and Schelling.[38] These (proto-Romantic and) Romantic thinkers paved the way culturally and ideologically for what would come later scientifically in terms of an evolutionary framework for understanding nature.[39] In other words, these thinkers were developing an understanding of nature as creative process, as *natura-naturans* or "nature naturing."

These Romantic thinkers of the eighteenth and nineteenth centuries were also influenced by Indic traditions. The influx of

Vedic traditions and their understandings of an active nature of which we are a part helped influence what would eventually become the science of evolution. Islamic and Arabic cultures had a huge influence on these Romantic figures as well, not just because a "scientific revolution" would have been impossible without their contributions, but because these Romantics drew directly from the Islamic world to develop their own thinking. Goethe, Schlegel, and Novalis, for instance, all engaged with Islamic texts and ideas. As Jeffrey Einboden argues, among the Romantics there is a "transnational arc of Western dependence on Muslim sources, stretching from Europe's earliest Romantic intimations, through the classic products of British Romantics, to the exhaustion of American Romanticism in the wake of the U.S. Civil War."[40] Thus at the heart of the development of earlier Romanticisms, we already see glimpses of a planetary tapestry: a recognition of the contributions of multiple voices and experiences of the worlds in which we live. These shifts in European understandings due to an influx of Indic and Arab/Islamic texts led to creative and different ways of understanding the natural world. Nature and culture are always "nature-cultures,"[41] and changes in one will affect the other. Haeckel and other scientists of his time drew from these perspectives and their relatively recent capacity to travel more easily around the planet in developing their own understandings of what some might call Romantic sciences.

By the time Charles Darwin wrote in the mid-nineteenth century, Haeckel was already convinced that humans were a part of the rest of the natural world. Darwin's theory simply made the origin of humanity from other species more scientifically plausible for Haeckel. He took up evolutionary theory with zeal and made it his goal to stamp out what he considered to be the ignorance promoted by the dualistic worldview coming out of monotheistic theologies. In his zeal and overconfidence, we might compare him today to a Richard Dawkins, and he invoked as much drama and employs as much hyperbole as Dawkins. Almost immediately in response to Darwin, Haeckel, and the formation of the Monistbund in Germany, which promoted the spread of evolutionary thought, a group of Protestant ministers known as the Keplerbund formed to stamp out evolutionary ideas and preserve a special place for humans in a nature created by (a Christian) God.[42] As Robert J. Richards notes,

> The most vocal opposition to evolutionary theory came from religious dogmatists. In particular, members of the Keplerbund (an organization of Protestant naturalists) objected to the antireligious and anti-Christian conclusions that Haeckel and others had drawn on the basis of evolutionary theory.[43]

It was the writings of the Keplerbund that captured the imagination of some in the emerging fundamentalist movement in the United States during the late nineteenth and early twentieth centuries and helped to lay the foundation for the culture wars in the United States during the twentieth century between creation and evolution.[44] Still today, if one searches "Ernst Haeckel" on the internet, one of the first results is likely to be a link to a creationist or intelligent design website disparaging Haeckel as a neo-Nazi, and rejecting evolutionary theory in general.[45]

Objections to evolutionary theory were not just from religious and theological folks, but also from those within the emerging natural sciences. One of Haeckel's mentors and friends, and then later enemies, was Rudolf Virchow, who helped found the field of anthropology. He opposed the idea of evolution of humans from other species, though not that of the process of evolution within species, largely because of the links he saw between evolution, eugenics, and racism.[46] Virchow wanted to preserve the "psychic unity" of all humanity, because he saw the racist implications of thinking that some "races" were more evolved than others. For Haeckel though, any sort of special creation of humanity by God ultimately led to some sort of dualism that he could not tolerate.

The battle against ideas of theological dualism was only one part of his three-front battle: he also fought against philosophies and methods that opted for strict idealism or materialism, like many Romantics before him. Idealism (especially of what he considered to be the Hegelian variety) suggested that whatever was really real, must transcend the material world: all reality is then reduced to spirit, energy, mind/consciousness, numbers, or forms. On the other side of the debate was materialism, which suggests that all reality can be reduced to chemicals, genes, or "matter" in general and that things like consciousness and ideas are merely white noise to the background reality of material and energy exchanges. For Haeckel, the wrong turn in German thought toward either idealism or materialism—though correct in their critiques of dualism—was to be found in the work of Immanuel Kant. He understood Kant's wrong turn to be located in his concept of "das Ding an sich" (or the thing in itself) and in his transcendent realm of a priori knowledge (or knowledge before experience).[47] For Haeckel both of these ideas ultimately sustained a dualistic worldview: a possibility for a totally other (realm) that was not accessible through the sensory experience of relationality and entanglement (a realm that cannot be attuned to). Such an unrelated "other" reinscribes a type of nonrelationality as the essence of identity. This continues a long line of thinking from ancient Greco-Roman to Western philosophers that suggest what is really real is transcendent or essential, and our differences are just fleeting material differences. Haeckel and other Romantic scientists wanted to return reality to the embodied world and with that to the focus on the diverse, evolving, entangled bodies in the world.

It is only later in the twentieth century, with the rise of postmodern discourses, that WM theorists begin to constructively think difference as really real. Of course, such thought would be nothing new for pantheists and animists, or the likes of theological mystics like Nicholas of Cusa,[48] but it was "new" for the dominant modes of Western thought that focused more on singularity and unity. Levinas's "otherness," Jacque Derrida's *différance*, and Judith Butler's idea of "abjections," for instance, embrace otherness, but not otherness as a transcendent unknowable essence. For Derrida and Butler, that which is other is a result of relation not before relation. Many Romantic scientists were hinting at similar insights, and in this way we might read a certain deconstructive or "queer" element in their thinking. In our relationships with others, there is always something that is deferred or left out of the relating, and these deferrals or leftovers (*differánce* or "abjections") come back to haunt the stabilization of relationships, according to Butler and Derrida. Eventually, this deferred otherness helps to break open ossified structures of relationships toward new ways of relating.[49] This is the type of reattunement process called for in a CPR for the earth, and we can find this type of thinking in nondominant forms of Romantic, Western thought (not to mention animisms and other immanent understandings of reality).

Romantic understandings of diversity and difference in nature are not a result of separation, not about anything a priori to experience or outside of relations, but of interconnection, and are thus akin to the later ideas of Derrida and Butler. Haeckel's triune monistic understanding of evolutionary and emergent diversity and difference meant also that all value emerges out of the natural world and all life therein participates in some way in the creation of said value. As I argue in chapter 4, aesthetics and values for many Romantics and definitely for the CPR I develop in this book, are emergent from the evolutionary process rather than transcendent to it. This move helps to trip up ideas of human mastery found in many dominant Western narratives. In general, I argue that placing humans in an evolutionary worldview provides the primary context for the emergence of Western postmodern and critical thought in the twentieth century. Evolution and ecology are the conceptual grounding for rethinking humans into the planetary community from a WM mindset. These sciences are the grounds for a "neo-animism" or a (not so) "new materialism."

In addition to critiques of the *ding an sich*, Haeckel argued against Kant's understanding of a priori principles and knowledge. For him these were not a priori claims, but rather emergent in the process of years and years of cultural and linguistic evolution, and the detritus left over from these processes.[50] Their historical construction was simply lost so that they seemed to be a priori. Such ideas edge close to postmodern understandings of "biopower." As Foucault would argue in the mid-twentieth century, those things which seem "natural" or "given" are in fact bio-historical processes that have, over time, coalesced into habits for becoming. They are co-constructed, but their co-construction is covered over and hidden as "nature."[51] As Donovan Schaefer concludes, these co-constructions include ideas, discourses, materials, and feelings; they include both the "seen" and "unseen." But the unseen is not transcendent nor out of the ongoing process of bodies reattuning. He writes that Foucault

> makes intelligible the shaping of fields of power by material culture. His approach washes out the binary frameworks that have structured so much thinking on power—not just the binary of the sovereign liberal subject set against external forces, but the binary of discourse and matter. Even speech is not just speech. It, too, has material-affective form. What Donna Haraway called the 'material-semiotic node'—an analytic lens that locates information and objects on a single interactive plane—can be productively reframed as the material-affective-semiotic node . . . . This is the theater of [Foucault's] *dispositif*.[52]

Though Haeckel was no Foucault, he did understand that there is nothing outside of this evolving matrix of energy, feeling, and matter and that whatever else ecology and evolution suggest, life is a constant process of reattuning to the entangled bodies that make up the earth at any given time.

If all knowledge is a posteriori, then there is a greater emphasis on the hermeneutical task of understanding the world, not because it's "true form" was somehow inaccessible to us before but rather because human reason is incapable of exhausting a world that is constantly changing. This "uncertainty" is simply the result of insisting that history is a part of evolution and geology (which is what Darwin's and Haeckel's, among others, conclusions implied).[53] It is from this perspective, one might argue, that placing humans within an evolutionary framework also places history in nature and is thus a necessary condition for postmodern questions of knowledge construction and hermeneutics.

Standard interpretations of Hegelian idealism and Marxist materialism would, then, be too transcendent or too reductionist, because they don't take all of reality fully into account. On the side of idealism, the suggestion that the really real might be located in some transcendent immaterial realm (spirit, language, numbers, ideas) is just a continuation of the more problematic side of supernatural understandings of the world promoted by some monotheisms. On the materialistic side, reducing everything to material does little to account for the reality of energy/spirit/language/mind, and furthermore suggests that humans can somehow transcend our own bio-historical contexts to "get to" or "objectively view" the bare facts of the world. How, one might wonder, would a purely material organism be able to reason or think about a purely material world? Wouldn't that very reasoning and thought be just reduced to chemicals or some other sub strata and thus not really real at all? If thought (feeling-become interiority-become thought) is really real, as are matter and energy, and none can be reduced to the other, then reality must in some sense be forever interpreted. Hence the need for a CPR, a way of thinking that takes natural and cultural location and embodiment, and the rest of the nonhuman world as agential, seriously.

One cannot get to a "base" level that does not require interpretation and is merely given. In this sense epistemological uncertainty is written into the world because of the very nature of the structure of the world. The nature of knowledge does not adhere to Heisenberg's epistemological uncertainty, but rather to Bohr's indeterminacy. Philosopher of science and particle physicist Karen Barad articulates the differences between these two positions well.[54] She argues that it is not just that we don't have the correct interpretation of the world and one day we will (Heisenberg), but rather interpretations will always remain interpretations and the worlds will be co-constructed differently based (at least in part) upon those interpretations (Bohr). This means we always need critical interpretations of the shifting and entangled worlds in which we live. For this reason, I think many nonreductive Romantic scientists and thinkers of the nineteenth century should be thought of as proto–critical romantics. Critical romanticism, like speculative realism, understands our embeddedness within a planetary community in a "real" world outside of the human mind, but it does not argue that we can ever attain a definitive understanding of the world or ourselves therein. Part of the reason is that we are embedded in worlds and cannot get outside of them to see them objectively. Another part is that the worlds we live in are evolving and open processes and are thus always changing, as are the knowledge regimes we use to understand the worlds from within. Thus, there is a need for constant reattunement to our changing worlds.

Though Haeckel and others like him were sometimes critically romantic, they often held on to the idea that one day all mysteries would be solved (certainty). This is a very uncritical stance. The belief in scientific progress and certainty (on the part of Haeckel, for example) seem to be at odds with critical epistemological and ontological understandings of the world. It is important to keep this in mind throughout this book. Rather than making the Romantics of the nineteenth century coherent "individuals" here, I want to keep their different selves in sight, contradictory though they at times they may be. Just as each one of us is a mixture of identities and selves over time, so are individuals of the past. Haeckel's triune monism provides a good example of the tensions that existed with these sometimes critical and other times naive understandings of Romanticism in the nineteenth century, and these contradictions provide glimpses into the societal and global changes happening at that time.

Haeckel's triune monism, which draws from the visions of Erasmus and Charles Darwin, von Humboldt, Schelling, Goethe, and Herder, was his description of the basic substance/structure of the world in which there is an emergent continuum from abiotic to biotic life, and eventually to conscious life.[55] The process of moving from the abiotic to the biotic follows the story, not of a creator god, but of the spontaneous generation of monera, for Haeckel. This spontaneous generation or "poiesis" involved the creative transition from nonlife to life. These monera "arose from the primeval ocean through the operation of purely physical and chemical conditions and were the origin of all living forms."[56]

Monera (coined by Haeckel) are single-celled organisms lacking a nucleus, made by an energy event, which brought together an "albuminous combinations of carbon."[57] In Haeckel's understanding, they contained merely one particle of protoplasm, which Haeckel believed was the source of what we might call an internally driven "life." Protoplasm is the soul/energy that is found in all living things. Though even in abiotic life, Haeckel argued that there is continuity with biotic life: as his book *Kristallseelen* (Crystal Souls) toward the end of his life makes clear. For Haeckel the distinction is between self-organizing life (biotic) and life that is acted upon purely by external factors (abiotic).[58] This is quite similar to what process thinker Alfred North Whitehead would later call democratic societies of entities vs. monarchical societies of entities.[59] Whitehead spoke of the attunement of all actual entities to one another, to the past, and to the future becoming of all entities in similar ways that Haeckel thinks about ecologies of organisms attuning to one another.[60] This is also similar to the idea of "ecological attunement" and has hints of what Harmut Rosa (another, more contemporary, scholar from Jena) identifies as "resonance sociology," which suggests individuals are formed by the interrelationships and structures that promote certain ways of becoming over others.[61] Finally, Haeckel's ideas are in line with discussions of contemporary emergence theory, which will be discussed

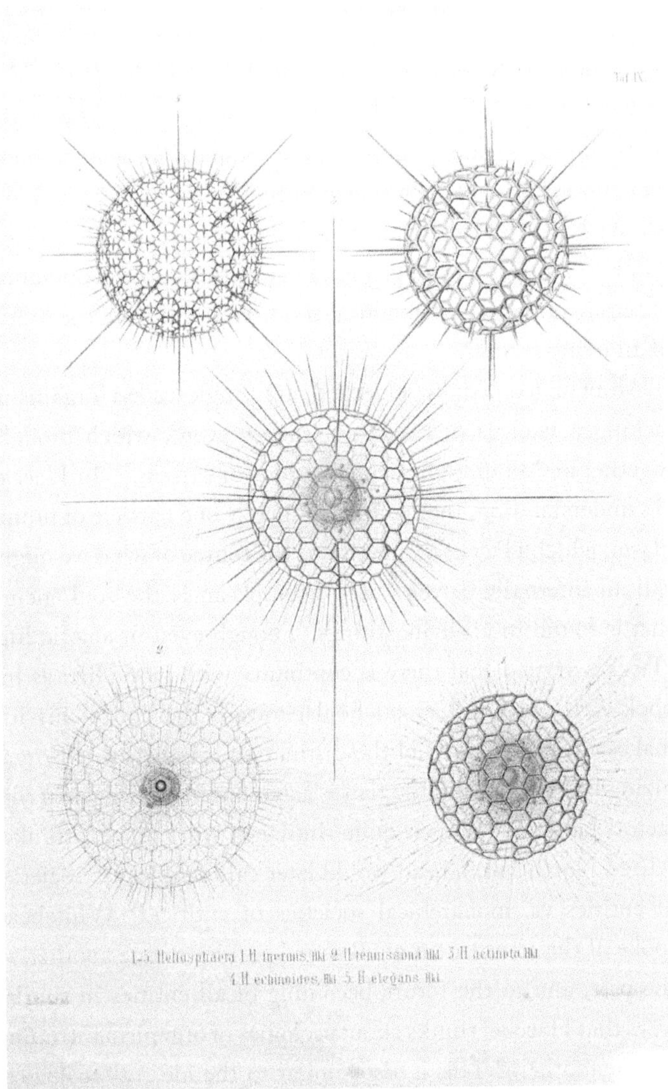

FIGURE 0.1 Ernst Haeckel, "Radiolaria" (plate 9), in *Die Radiolarien (Rhizopoda radiaria): Eine Monographie,* vol. 1, *Atlas* (G. Reimer, 1862), www.biodiversitylibrary.org/page/10687513. Monera were the predecessors to radiolaria, single-celled organisms. Throughout Haeckel's artwork, he wanted to show the continuity from abiotic to biotic life, plant and animal life, and human life and the rest of the natural world. Monera represented the shift from nonlife to life.

further below. Indeed, for Haeckel the concept of ecology is one that is necessary to explain evolutionary differences: these differences are found in an individual's "attunement" to the community of life it is surrounded by.

Before the formation of monera, the precursor to biotic life were crystalline structures, which is one reason Haeckel pays so much attention to the patterns that make up crystalline and abiotic structures. For him these patterns are also found in biotic life, except with biotic life each individual organism begins to "remember" (or attune to) the world—via the protoplasm—and pass those "memories" on to their progeny (the beginnings of interiority). The transition from abiotic life to biotic (ensouled) life, then, is more of a transition of the locus of organization of a life than anything else. In other words, whereas rocks and crystals and other forms of abiotic life are "shaped" (or attuned) by the environment and climate around them (what Whitehead called a democratic society), beginning with Monera entities begin to react to and interact with their surroundings, a process of intentional attunement.[62] With this comes an ever more complex internal locus of control, eventually leading up to animals and humans (which Whitehead called monarchical societies).[63] Of course, even self-organizing, complex creatures are never fully "in control." Everything is acted upon and acts upon other things in ways that shape all entities becoming. All of this means, among other things, that life is continuous with nonlife—indeed life is an interiorization of nonlife—and that physics, chemistry, geology, and biology can all begin to fit together into a planetary story: a story in which life, energy, feeling, and eventually thought and consciousness can be understood better within an immanent framework. Within this framework various entities are created through a process of attuning to surrounding entities.

thinking. And, though many theologians at the time were still thinking in terms of a dualistic and transcendent understanding of value, truth, and beauty, there were a growing number of exceptions to the rule. This was part of the overall shift toward an immanent frame of reference from that of a transcendent one.[66]

This shift is still not palatable for many Western moderns today. For Charles Taylor, the shift to an immanent frame is one that is a collapse into materialism and meaninglessness. However, this collapse into meaninglessness is only "necessary" if one sees the shift to immanence as reductive and not as something which opens possibilities for becoming. Such meaningless materialism is only possible if value and agency reside solely in a transcendent realm. If one gets rid of the transcendent (killing off God), then meaning is killed with it, leaving only pointless materialism: life is just one damn thing after another. However, if meaning pervades everything then this is not the case. As Bruno Latour argues, once you shift value, purpose, and meaning from something anchored in a transcendent reality, then the world has many more possibilities for how it might become.[67] Such a collapse of transcendence, then, opens us on to an open-ended, emergent form of immanence. A form of immanence in which there is an open-ended planetary evolution of emergent possibilities that must be constantly reattuned to.

As mentioned above, the Protestant Keplerbund spent a lot of time combating the ideas of Ernst Haeckel and other immanent thinkers grounded in some type of evolutionary thought in general. At the same time, they were concerned with Marxism and its threat to the authority of the church, not to mention the shifts in class structures proposed by Marxism. Many philosophers and social scientists argued that Darwinian thought was the foundation for a Marxist view of the world (or vice versa), and thus pitted Marxism and Darwinian evolution against a theistic understanding of the world. As Daniel Gasman points out in his problematic assessment of Haeckel, the horrors of World War II helped to create a connection between Darwinian thought, Marxism, atheism, and the rise of eugenics and Nazism, on the one hand, fighting against the good, creation-based, democratic and theistic views of the world, on the other.[68] Mind you this democracy (in what is now the United States and elsewhere) was built on slavery and genocide, and the subsequent inequalities resulting from that, but these facts are conveniently ignored in this creationist vs. evolutionist debate. This basic division helped to wedge a divide between religion and science in WM cultures that still exists for many to this day, most notably found in the type of "two worlds" mentality that suggests science deals in reason/facts whereas religion deals in faith/values, and even the continuing-warfare model where science and religion are fighting an epic battle. The corresponding methodological grounds became something like: science/reductive materialism, religion/idealism. For many scientists and philosophers of the nineteenth century, science and religion were not so distinct; they were still part of a single tapestry or worldview. Indeed, for Haeckel and his own problematic evolutionary understanding of "religion," monism would be the new religion based upon evolutionary theory that would replace all other religions. However, in retrospect these subtleties have been abandoned for an anachronistic understanding of history that reads "science and religion deal with different realms" or "science vs. religion."

These political/historical battles were only part of the reason for the methodological slide into reductive/productive materialism on the part of the natural sciences in the twentieth century. There were also strong economic and political reasons. The Industrial Revolution had begun to show the economic benefits of a reductive and productionist model. Science as a method of producing wonder about the world and exploring the world for understanding, such as the type that Haeckel and other Romantic scientists promoted, was all well and good; but the science of economic production and technology transfer was what national governments, pragmatic socialists, and capitalists all favored. Furthermore, making all nonhuman nature and even some humans more like "nature" was a politically expedient way to justify the use of nonhuman nature toward the idea of progress in history for colonizers and settlers, as well as the enslavement, education, and civilization of humans deemed by the colonizers to be not fully human. This, of course, included many Indigenous peoples, Black and brown peoples, women, and even children.

Up until the time of the two world wars, there was still a space for the science of exploration and wonder, but the wars helped to transform science toward a "productionist model":[69] first toward the production of better weapons and war technologies, and then later toward the production of better modes of production, communication, transportation, and medicine.[70] Again, these technologies would benefit some of humanity more than others, and still treated the rest of the natural world as resource for (some) human's progress. The economic benefits of such technologies became a feedback loop for keeping sciences tied to this productionist model: more money was (and is) available to fund research that lead (and leads) to results that could (and can) be translated into technology transfer. There were of course many scientists then and now who argue(d) against this reductive, productive model, but the economics of it tends to favor the technology transfer of the reductive, productive model, and the disciplinary boundaries that keep creating the world in their own reductive and productive images.

It is important to note that this reductive productive model was also employed by the Nazi regime: the reduction of everything to resources for human projects. Of course, the problem then, as it is still today, is deciding who/what counts as a human with rights. Just as the Nazi regime had an ideal human person to which all other life was subordinated as resource for, so the reductive productive model of science in the post–World War II modern West improved the lives of some at the expense of others. Nonhuman animals (persons), the environment, and the labor and bodies of "poor peoples of color" often became fodder for the improvement of life in wealthier nations. In the end, I argue that though one can link German Romanticism with a certain sense of nationalism and wholism, and thus as part of the fuel for a Nazi ideology, it was the reductive and productive model that caused much more violence.[71] Attaching Romanticism to a planetary model and to critical theories, as I will argue, guards against this type of nationalistic violence.[72]

The "productionist model" finds its ground in something like a reductive materialism, which narrows reason to mere instrumental reason (forget practical reason not to mention something like wisdom) and causality to mere efficient causality (forget formal, material, and final causality). The problem with this, of course, is that the world is much more complex than instrumental reason and efficient causality can account for. As critical theorists Max Horkheimer and Theodor Adorno warned, if we try to press the world into the confines of human (instrumental) reason, we will end up destroying the world and humans therein.[73] Though the horrors of World War II and the threat of nuclear annihilation eventually helped to crack open belief and faith in the narrative of scientific progress, it was Rachel Carson and other early environmentalists like her in the

1960s that began to seriously question the reductive/productive model for science from within the sciences.[74]

It wasn't really until Rachel Carson published *Silent Spring* that the reductive and productive model of science was seriously challenged in a sweeping way.[75] It was this book, more than any other that set off the WM environmental movement in terms of thinking in a more integral way about how human productions affects ecosystems and even our own bodies. Though some point to the conservation and preservation movements as a beginning of WM environmentalism, I would argue that those were just continuation of colonial projects more than anything else. It was really with Carson bringing awareness to how chemicals and toxins were destroying the water, air, and land that make up our planet and all planetary bodies therein that started an integral environmental movement which understood humans as a part of nature. What Carson realized, in a sense, was the material realization of Horkheimer and Adorno's warning about reducing the world to instrumental human reason alone. This was a similar warning that Heidegger was making about human "enframement" of the world and turning it into things (standing reserve) vs. understanding the world as autopoietic, as valuable in itself, and allowing nature to nature.[76] The former will end up in the death of nature and eventually humanity; the latter makes for a more humane, ecologically sound, and pluralistic planetary future. Carson's own research realized the truth of these earlier philosophical assumptions (even if not intentionally) in *Silent Spring*.

It was Carson's ability to "unlearn" the dominant voices of the sciences she knew in order to understand the connection between pesticides and their effects on other life in the world, and retrace the connections between chemicals, the thickness of bird shells, and even cancers in a more complex way, that allowed her to come to her major conclusions. In other words, she had to become undisciplined in a sense, thinking outside of the confines of the disciplines of her era so that she might (re) attune to human-earth relations in a new way.[77] Most of the (male) scientists during her time criticized her deeply and tried to discredit her ideas in *Silent Spring*. Many of these scientists were, of course, already receiving funding from the industries producing the materials she critiqued. Though Carson did not develop any sort of wholism or nonreductive materialism, she was very much influenced by Albert Schweitzer's "reverence for life"; she developed a partial "sea ethic" or "watery aesthetic" based upon the mystery and wonder of the natural world; and she, like the Romantics before her, argued for the benefits of direct study of the natural world, one of which was that such study led to a sense of wonder.[78] This will be discussed further in the final chapter of this book.[79] Science for Carson was not divorced from some type of meaning-making practice.

Now, many philosophers and scientists, environmentalists, and others concerned about planetary climate weirding, the sixth mass extinction period, environmental injustice, and other forms of water, air, and land pollution are beginning to question seriously (as Carson did before them) the reductive/productive model of the natural world. Some argue for "wholism" while others argue for a type of nondualistic, immanent, pantheism or panentheism, and still others for a nested emergent reality with multiple levels that are all interconnected. None of these ideas are foreign to discussions that the earlier Romantics were having. I am aware that contemporary models for thinking immanence have differences from their ancestral models and from one another, but nonetheless, they are all part of the same family of ideas that shift away from thinking dualistically and reductively in a way that honors both the material and ideal as real. This book is, in the end, an attempt to bring earlier Romantic scientists into discussion with some contemporary forms of nonreductive thinking; show that reductive and productive models of science are more of an aberration than the norm; and warn all those seeking to ground values, knowledge, and truth in "nature" of the dangers of doing so naively. In short, I want to critically retrieve Romantic scientific thought as a support for nonreductive models of nature in order to develop what I am calling a critical planetary romanticism or CPR for the earth.

CHAPTER-BY-CHAPTER ANALYSIS

Though reading this book all the way through is the "best" method in my mind, I have organized it so that you can read this longer introductory chapter and then skip around to the later chapters that interest you the most. Each chapter starts and ends with a bit of a recap and a description of where the next chapter will go. So, for those pressed by fossil-fueled time (aren't we all?), I have tried to make it easier for you to get the "gist" of the book in this introductory chapter and through recaps in subsequent chapters.

Chapter 1 turns to the implications of ecology and evolution for understanding radical immanence and relationality. My argument there is that placing humans and all things human within an evolutionary and ecological framework, as part of the rest of the natural world, changes how we source and ground knowledge claims. Haeckel and other Romantics spent much time reattuning all things human to Erasmus's and Charles Darwin's (and others') insights about nature. History, language, and culture become "emergent" in the process of geo-evolution rather than being somehow transcendent to the rest of the natural world or serving as foundational claims into which the rest of the natural world must fit. Furthermore, placing humans within an evolutionary context means that all humans interpret the world from within embodied, bio-historical, natural and cultural perspectives. All knowledge of the world, then, must wrestle with the problem of hermeneutics: put another way, everything is interpreted. This doesn't mean that there are not better and worse interpretations, but it does mean that there are multiple perspectives on the world that cannot be reduced to a single, objective truth. In short, placing humans in an evolutionary context is a necessary condition for the rise of postmodern discourses: to know anything, humans must not "go far away" into an ideal space but rather must reattune to the grounds of the worlds around them.

In chapter 2, I focus on some of the theological and religious influences on scientific understandings of nature. Haeckel's monism, for instance, was triune: it included matter, energy, and sensation or feeling as three different aspects of the same substance. It was also, unlike Spinoza's pantheism and more like Herder's, Schelling's, and Goethe's, an open form of monism rather than a closed form: it allowed for evolution and new things to emerge in the process of nature naturing.[80] In this sense, Haeckel's understanding of a naturalistic worldview—drawing from theological understandings of progress, German Romanticism, a critique of Kant, and even a Schleiermachian turn to experience and embodiment—offers a different understanding of humans, animals and the rest of the natural world than the model of productive/reductive materialism under which much scientific work is "done" today. Furthermore, because all the Romantic scholars were drawing from Indic and Islamic/Arab literature (especially through Novalis and Goethe) to develop their ideas about "nature," Haeckel's very understanding of evolution and ecology, and all things human therein, is only possible through a "planetary moment." The Romantic moment, then, was a planetary moment fueled by the exchange of resources, energy and information, and the technologies of travel that made visiting places around the planet easier; it provides one (but not the only) early example of a proto-CPR.

In chapter 3, I move to a discussion of Romantic understandings of "aesthetics from below." Just as the evolutionary context changes ideas of knowledge and what we can know, so too does it change our ideas about beauty, affects, and forms. Romanticism shifts the focus of beauty and forms from that of a transcendent realm, which is imposed upon the material world, toward that of immanent, emergent forms, with which we humans have evolved along with in a way that we can appreciate and identify through deep and direct study of the natural world. The emergence of forms leads to a breaking open of what it means to be male, female, human, animal, plant, and mineral in a way that opens the human and more than human world to a greater diversity of expression and interpretation. Nature is, we might say, queer. In the end, however, Haeckel's Romantic understanding of diversity was still plagued by the Eurocentric, nationalistic, and monotheistic progressive thought of the time in that all evolutionary life is ranked in a value hierarchy, the apex of which is Europe (and German culture in particular).

This latter, more tragic point is where chapter 3 ends and chapter 4 begins. As much as earlier Romantics might reimagine beauty and forms from below, there is also a tragic side of nature, and a tragic side to Romantic thought. Life only exists at the expense of other life. That death is a part of nature naturing and a planet made up of bodies that are emergent, entangled, and evolving means we all get eaten in the end. The response to this can't only be the transhumanist fleeing of our bodies and the planet as this just perpetuates the most problematic aspects of a reductive, productive science.

The metaphors of the light that science shines upon the natural world, coupled with the metaphors of the darkness of superstition and religion, and the dark, tragic side of nature, helped to fuel a racist scientific project that promised to bring everyone out of darkness into the light and correct the mistakes of nature. Development and progress for some has been underdevelopment and regress for many others (both human and nonhuman). Beauty depends upon the recognition of that which one finds ugly. Anyone developing a naturalistic worldview should be prepared to discuss the costs of such a vision and identify who/what pays those costs. This chapter argues that ideas of progress across the political spectrum have created too much certainty and violence, and that we need to take some account of the violence inherent to any understanding of progress. This is why I argue some form of a CPR might serve us better.

Chapter 5, then, begins a more critical and constructive retrieval of Romanticism for a planetary context.[81] By planetary context, I mean one in which our daily lives are even more connected than a hundred years ago, due to the increases in the speed of communication, transportation, and production (the so-called great acceleration), and one in which we understand ourselves, through the sciences of ecology and evolution among others, as radically interrelated with all other life on the planet. National, and other parochial identities no longer hold up in this era of planetary entanglement (if they ever really did). Returns to a "smaller" world are not an option, nor is the continuation of neoliberal globalization as usual with environmental and social concerns tacked on like band aids. Instead, I think we need spaces of planetary polis, so that we can begin to imagine future options that take into account our current realities in more just and ecologically sound ways. My small offering is a CPR for the earth.

Chapter 7 concludes the book with an analysis of philosophers of science and environmental philosophers, critical theorists, and ethicists (not to mention other scholars of the humanities and scientists themselves) who remind us to be wary of the violence that certainty can cause. Whether the certainty of science or the certainty of religion, a certainty that tries to close off and contain life within it will end up suffocating emerging, evolving, entangled bodies. We need instead to allow for the "undisciplining" of the productive, reductive knowledge that has brought us the world of climate change, extinction, gross economic injustice and environmental and social injustices. We need what Mel Chen describes as "a turn to intoxicated method as a justified mode of (dis)engagement; it is a mode of unlearning in part precisely through the ways it allows metaphysical suspension or worldmaking already established."[82] Or as Freya Matthews suggests, we need to focus more on "encounters with" rather than "knowledge of" others.

Scholars concerned with the problems of reductive understandings of life argue that nonreductive models—emergence theories, new materialisms, queer theories, and affect theories among others—might provide a better understanding of the planetary community that humans are inextricably a part of, and in which there are many other actants and agents than humans. However, this can never be a naive "return" to some sort of depoliticized experience and attunement with the rest of the natural world. Rather, we must use critical theories to understand how different embodiments experience and attune to the worlds in which they live differently for better and worse. There is no apolitical space in rethinking the planetary beyond reductive/productive models. Thus, a CPR must always be involved in self-critique and must always understand that there is no single world, but rather many worlds worlding together which make up at any given moment what I am calling the planetary community. In the end, this is a process of reattuning to the worlds in which we live, critically examining how bodies (human and non) are affected differently by these worlds, and reworlding in ways that are inclusive of more and more planetary bodies, ad infinitum. The afterword provides the beginnings of an open conversation about some type of criteria for what ideas and worlds might be "better" or "worse" from the perspective of a CPR.

I

RETHINKING TRUTH (RELATIONALITY)

1

EVOLUTION AND ECOLOGY

The Sciences of Radical Immanence and Relationality

Science is often referred to as if it is a stable unchanging endeavor. However, as twentieth-century philosophers of science have argued (Karl Popper, Thomas Kuhn, and Sandra Harding, among many others), science is every bit as much of an interpretative endeavor as any discipline in the humanities. The cultural, historical, embodied location of the scientist matters in terms of what is studied and how data are accounted for in the meaningful narratives that make up our understandings of the worlds in which we live. Ernst Haeckel's work provides an excellent example of the importance of a hermeneutics of scientific understandings of nature. Haeckel argued for his monistic (nonreductive materialistic) understanding of the natural world against what he perceived to be the dogmatic slumber of theology. His embrace of Darwin had every bit as much to do with a rejection of traditional religious institutions as it did with adopting the "best" scientific theory. Here we have an example of how the territories of science and religion are mapped based upon their interaction throughout history: What we think of as science here and now would not be science without religion and vice versa.[1] This is the case in different times and different places. The "data" or "experiences" or both that are covered by a "map" of science or religion must take this historical-cultural location into account as part of the process of interpretation. Haeckel realized at some level this hermeneutic task and that the "best" scientific theory still involved interpreting data.

One epistemological challenge of evolution is that it places science (constructed over time by human-earth interactions) within an immanent frame with everything else that exists and thus turns it into an interpretive endeavor. It moves the objective eye from above to a located body within a process. Or, as Walter Mignolo argues, it transforms thinking from "I think, therefore I am" to "I am where I think."[2] Knowledge is not so much from the perspective of an engineer or architect, from a space outside in which the whole can be seen, but rather from the perspectives of the carpenters, painters, farmers, and other laborers within. It is, in this sense, relational. The "geologian" Thomas Berry recites a good metaphor for this type of work and experience. In *The Great Work*, he reminds us of the centuries that it took to build medieval cathedrals, the colosseum in Rome, or the Great Wall of China.[3] Most of the people who worked on these projects never got to see the final product, much less experience the purpose of the completed building, but they knew there was common work together to be done, for a greater purpose. This is a much better metaphor for knowledge production than knowledge as an unobstructed view from nowhere.

Ernst Haeckel showed signs of both detached and embedded understandings of knowledge. From an Enlightenment, objective perspective, he resorted to wild, inflammatory statements against his harshest critics, arguing for the certainty of science and the delusion of theology. In his more critical, reflective, philosophical moments, when he understood the implications of placing humans within an evolutionary perspective, he was more of a cobbler of ideas and knowledge from science, philosophy, and even (surprisingly) religion in an effort to make sense of the natural world (and humans there in). This chapter takes a closer look at the scientific influences on his work and ideas. Though science and religion can never be fully separated, they do require different organizing principles and foci. Science and religion, to extend Timothy Morton's metaphor, function more like a Mobius strip: in the end science is science and religion is religion, the laboratory is a laboratory, and the church, mosque, synagogue, or temple is a church, mosque, synagogue or temple; yet they cannot be fully separated.[4]

It is important to note here that Haeckel is writing during the process of the "discovery" of historicity. On the one hand, due to the increase in travel and interaction with "other" cultures and religions, you had anthropology, history of religions, philology, and other humanities asking both how one culture/language related to another, and what the "origin" of humanity, human languages, human cultures and human races might be. This led (not inevitably) to the ranking of differences according to a progressive or developmental understanding of time, which meant some religions, languages, and cultures were seen as "more advanced" than others. On the other hand, you had evolutionary theories, geology, and the beginnings of cobbling together a fossil record, which led to an understanding of the historicity of the natural world into which Haeckel (following Erasmus and Charles Darwin) would try to fit humans and all things human. To paint in broad strokes, you had the evolutionary spirit/history of Hegel and Marx, and the evolutionary material of the Darwin's (and others') coming together to help foment the turn towards historicity (and geohistory)[5] in the eighteenth and nineteenth centuries. This turn would help to create the promises and perils of pluralism and postmodernity that persist today. For Haeckel, this turn would be most influenced by Charles Darwin's publications, though many other evolutionary thinkers had come before.

Whereas an immanent aesthetics was central to Haeckel's way of understanding and knowing the world (and central to the scientific endeavor in general), religion was for Haeckel a secondary characteristic, so to speak. In other words, he drew all of his authority from what he interpreted to be "the scientific worldview," and then figured out how to fit religion and ethics into that worldview or at least asked how religion made sense according to the sciences (of course, he misses that this is what many theologians following natural philosophy had been doing for centuries). As such, he, like many other natural scientists of that time, had a bit too much confidence in what science can tell us about truth and knowledge, and was a bit too eager to destroy other ways of thinking and understanding that fell outside of the scientific worldview. Furthermore, his epistemological/methodological preferences also perpetuate the separation of science and religion/theology, and nature and culture more broadly. If all things human are part of nature, then our ideas, values, cultures, religions, and even sciences, ought also be part of the data interpreted, and from which we draw conclusions.

Haeckel and others of his era didn't fully realize the things they imported from theology and philosophy, their own subjective-historical-cultural locations within Europe dominated largely by Christian backgrounds, even though many other cultural-religious influences existed and persisted, heavily influencing their readings of the natural world and what they constituted as "scientific." Whereas Haeckel and others did recognize that all knowledge was interpreted, many of them didn't think that anything outside of the emerging "scientific" disciplines had much different to offer by way of common knowledge. I'm not suggesting that some ways of thinking don't need to be challenged, nor that there are not better and worse types of knowledge for understanding the world; but I am suggesting that there is an evolution and ecology to knowledge production that Haeckel seemed to lack full awareness of. Ideas emerge in and out of contexts, not ex nihilo.

I argue that overconfidence by some scholars of science in the nineteenth and twentieth centuries likely stems in part from the overzealousness many scientists had in stamping out theological dualisms, and from the material and economic benefits of the reductive and productive model of science. This overzealousness, however, is only created through an anemic understanding of how religious people were and were not engaging in science at that time. First, much of science was still being funded by religious organizations, and many scientists were also students of theology, if not clerics. Even Darwin himself, of course, was a trained theologian. Second, there were whole groups of clerics who were also naturalists and instrumental in spreading newly emerging scientific information.[6] We might say there were a lot of "religious" scientists, and a lot of "science" and the spreading of knowledge created by it was still done because of religious initiatives. Third, in creating an "anti-" position to dogmatic theology, the structure of authority that some scientists were fighting against was simply mirrored. Whereas priests and the church once had authority in metaphysics, science would now have authority in all things: from educational reform, to aesthetics, to "spiritual" matters and the ethics surrounding life and death. Haeckel's zeal for arguing against theology, for instance, and his inflammatory comments about the God of Christianity as nothing more than a "a gaseous vertebrate" brought him scorn not just from the church and some Protestant theologians but also from Charles Darwin himself. Indeed, in one letter from Darwin to Haeckel, Darwin begged Haeckel to tone down his rhetoric against theology so that people would not be so resistant to his evolutionary ideas.[7] In other words, Darwin knew that the deeper implications of evolutionary theory might fall on deaf ears if those attached to those ears felt attacked, so it was better to stick with the more strictly "scientific" side of things and just leave theological questions unanswered.

It was in no small part the rhetoric of scientists such as Haeckel and Huxley that led to huge rifts between evolutionary theory and those that positioned themselves against these theories—the latter of whom who would later become known as creationists of one form or another. This constructed rift between an evolutionary understanding of the world and a religious understanding of the world has had horrible consequences that will be discussed in later chapters. Here, I turn to some of the implications of placing humans within an evolutionary context, within a "history of the earth."

THE EVOLUTIONARY CHALLENGE: THE ETHICS, AESTHETICS, AND POLITICS OF *NATURA-NATURANS*

During the Glorious Revolution, John Locke and Robert Filmer had an infamous debate over the interpretation of the Genesis 1 account of creation. The Genesis 1 narrative with its six days of creation was, for the most part at that time in the Western world, one of the more popular stories of creation, even if "days" were understood as metaphorical units of time.[8] Political philosophers in what is now Europe and the United States would look to this story as the story that grounded understandings of human identity and organization. For Locke, the idea of the *imago Dei*, and the fact that dominion was given to the first representative humans, meant that all humans were given dominion equally. Thus, some form of democracy, not monarchy, was the appropriate form of governance. For Filmer, on the other hand, the original two given dominion were symbolic of monarchical rule.[9] Thus Locke and Filmer had two totally different interpretations of creation, and the location of humans and forms of governance within.

The emerging theory of evolution, and the place of humans within it, also set off multiple interpretations. This time, however, the crisis included one of shifting authority about the order of creation from that of religious experts and texts toward those of "scientific" ones. Haeckel and other Romantic scientists represented the thinking that nature naturing was the primary context for humans within the ongoing process of evolution. In fact, he understood Darwin to support him (and he did to some degree) in thinking that species are mere nomenclature for an ongoing process: in reality, there is only ongoing nature or the transformation of energy, matter, and "feeling." If humans are placed into such a nature, then it means that the dominant human exceptionalism proposed by many religious traditions, not to mention most European philosophical traditions, but especially those of the monotheistic Western variety, must be drastically rethought.

Whereas some thought that humans were just another part of the process, though perhaps the most highly evolved (Haeckel), others still clung on to the idea that there was a single source of evolution within the human species (Virchow).[10] Part of the reason some early anthropologists such as Virchow held on to a special place in evolution for human beings, for instance, had to do with the racist implications of ordering species from more primitive animals to the most "advanced" humans. If everything "evolved" from a common origin and there was progressive time, some scholars assumed that differences must be ranked hierarchically. This meant that some humans might be seen as "closer" to animals and as "less" evolved. This progressive and singular time was and is, of course, an abstraction that in reality forces the entire planet to conform to a single tunnel of time. It leads to the interpreter's own location being understood as the "most progressed" and all other things and organisms are ranked accordingly. Haeckel made the mistake of ranking the races in this way, whereas Virchow argued that within humans there was a "psychic unity" and that all humans might be seen as just different manifestations of what it meant to be "human," without the racial-cultural progressive hierarchy of valuing humans above animals and some humans above others.[11] In other words, for Virchow holding onto some sort of original human form was a way of claiming equality among humans.

There were other battles over what placing humans into an evolutionary context meant for economics and politics. Some thinkers read capitalist and aristocratic structures into the hierarchy of an evolutionary process based upon "survival of the fittest."[12] Still others were arguing that evolutionary theory led to a more Marxist understanding of politics and economics in which some sort of socialist and maybe even communist structure would emerge as the highest form of human organization.[13] Others, such as Haeckel, refused to tie evolutionary theory too closely to a political and economic ideology even though he (subconsciously?) leaned toward a progressive, aristocratic interpretation, and understood his own interpretation as "neutral" and "natural." He argued that humans were, evolutionary speaking, of the same origin as dogs, while at the same time arguing for the superiority of certain races over others: which

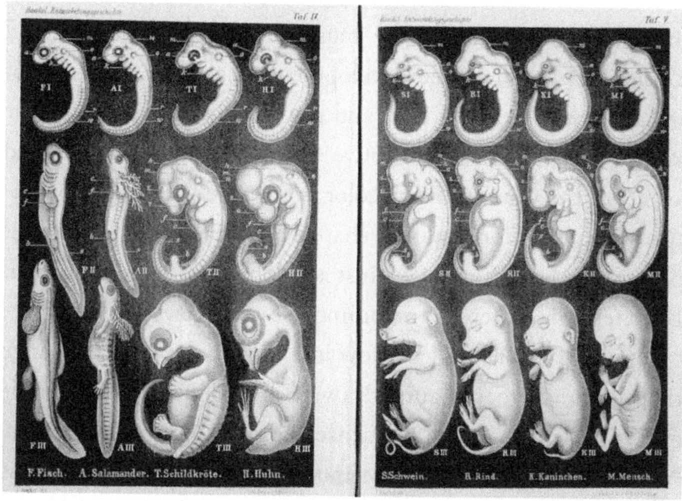

FIGURE 1.1 Double-plate illustration showing embryos of fish (F), salamander (A), turtle (T), chick (H), pig (S), cow (R), rabbit (K), and human (M), at "very early," "somewhat later" and "still later" stages, from Haeckel's *Anthropogenie*, published in 1874.

could be evened out through a process of ongoing education of the "lesser" races.[14] Ideas of race, class, and politics were read into nature, and were also challenged by new readings of nature, just as had been the case in previous eras.

Haeckel, like many others, mistook his own subjective position as somehow neutral; or rather as a scientist he, like many scientists, failed to recognize his own socio-historical context. The fact that Haeckel spilled so much ink arguing that everything human is part of an immanent, evolving nature and thus all facts must be interpreted, and that he still had huge blind spots in terms of how his own socio-historical location shaped his own interpretation of reality, should be a huge warning or reminder for all of us interpreting nature from any location. No matter how much we control for biases, there is always an interpretive element in the scientific method. This is the reason that any interpretation of nature must be a critical one, which I argue is essential in the development of a CPR. Goethe, for instance, was aware of this, and thus in his theory of color he argues that there is a perceptive, physiological, and embodied aspect of the ways in which color is perceived. The eye apparatus of humans, for instance, shapes the ways in which color is perceived, thereby going against any "objective" theory of color. By extension, different embodiments would have different perspectives of color.[15] Indeed now it is quite common knowledge that different types of animals (with different types of embodiments) hear, see, smell and sense the world in different ways. Nature is, in this sense, always multiperspectival, so there ought to be a plurality of (partial) interpretations of it from different locations. After agreeing to evolutionary theory, there are no more external, "safe," or objective places from which to judge nature in its entirety. Embracing a hermeneutics of evolution means embracing the postmodern condition.

HISTORY IN NATURE: BIO-HISTORIES AND NATURE-CULTURES

My argument in this section is not that Haeckel himself, nor any other nineteenth-century scientist of his time, was "postmodern." Though Haeckel was beginning to articulate what we might now call a postmodern epistemology, he was still located firmly in a modern, Enlightenment mindset. Rather, my argument is that some of the very tenants of postmodernity—the problem of historicity, all reality is interpreted, identities are nonessential and fluid, and values are co-constructed over time—depend in large part upon humans being placed within an evolutionary context. By evolutionary context, I simply mean that humans are not exceptional to, but rather of, from, and with the rest of the natural world. This doesn't necessarily have to be based on a scientific understanding of evolution; it could also be based on a more relational or immanent understanding of humans as part of the rest of the natural world, such as found in pantheisms or animisms.[16] If human identities, values, and histories emerge out of a common planetary context with all other life, then human knowledge of the world, and human cultures and technologies, are also emergent and contextual: knowledge is relational. This means that as our evolutionary context continues to shift and move, we must reattune to the newly emerging worlds around us. Knowledge is a process that is never complete or final. Haeckel, at least in his more philosophical moments, understood this emergent nature of human knowledge, identities, histories and values in several ways.

First, in his more philosophical moments Haeckel argued that there are no bare facts in nature, rather that all understandings of nature involve interpretation. Put in philosophical terms, all knowledge is a posteriori knowledge. For Haeckel, the fact that all knowledge comes through experience and empirical evidence made developing a naturalistic worldview in which all peoples were charged with the duty to study nature for themselves all the more important. He was flustered by the fragmentation between the emerging chemical, biological, and physical sciences in the nineteenth century, and understood evolutionary theory as a way to finally bring all of these multiple perspectives on "nature" together into a single worldview. He called this worldview monism. For Haeckel multiple perspectives from the various sciences would give us a much better understanding of the world in which we live than any single science could alone. The different mappings, from the different sciences, could give us a more in-depth picture of any given phenomenon precisely because more knowledge is produced when more perspectives examine phenomena. These different perspectives included both the natural and human sciences, even if the evolutionary framework was the all-encompassing one and the human sciences ought always to give final authority to the natural sciences.[17]

Despite his understanding of the hermeneutic necessity of interpreting the natural world, he still hoped (from his embeddedness in enlightened and modern European culture) that this naturalistic worldview would provide more certainty in understanding nature where theological certainty had faded. In other words: he (like other Westerners in the wake of the Scientific Revolution and Enlightenment) replaced theological certainty in a creator God's divinely revealed creation, with that of a Nature whose laws were discoverable through observation and reason. For Haeckel, knowledge of the world and our interrelatedness with it, could only come about with the help of a nature aesthetics in which we could witness the evolution and emergence of forms, rather than assume that forms were imposed upon nature ex nihilo in some cosmic beginning.[18] For Haeckel organisms in the world (including humans) emerge from and together in such a way that the natural is the source of the beautiful: that is, aesthetics are derived from nature naturing.[19]

Second, and related, loosing forms from transcendent foundations (in a creator God or some other form fixing entity) meant that the world was much more diverse than humans could previously imagine. Placing humans in an evolutionary context was more than just rethinking what it means to be human vis-à-vis other life on the planet, or what this meant for one's religious understanding of the world, as important as these things were and are; it also meant tracing how things like thoughts, ideas, values, and consciousness emerged from a long line of evolutionary developments. Rather than a top-down approach to knowledge, this new way of thinking about the world and what it meant to be human required a bottom-up way of thinking about the world, and with that an understanding that the forms once thought to be somewhat stable change over time: What does this mean for "the human"? What does this mean for distinctions such as "male" and "female"? What does this mean for "beauty"? These are the types of questions that in an evolutionary context require new answers, and Haeckel and other scientists of his era realized this. Part of his vehement opposition to theological dualisms, idealisms, and reductive materialisms, was that he understood these as old ways of thinking from within a pre-evolutionary context. Much like Alfred North Whitehead needed new language in developing "process thought" based upon his interpretation of new physics, so Haeckel needed new language to talk about this new, emergent and relational understanding of reality, and Darwin, among others, supplied him with such a language.

Many pre-evolutionary, Western ways of thinking relied on something like Foucault's panopticon or all-seeing objectivity. Of course, there were immanent traditions within Western thought from at least (if not always) Heraclitus on, and Haeckel draws from these, but the dominant Western model for knowledge was based on transcendence, unity, and stability. A framework in which we could, through reason, see the whole picture of life from a "god's eye" view, was made nearly impossible by this shift to an evolutionary, immanent frame that places humans and cultures into nature. I say almost impossible because many philosophers, theologians, and scientists persisted with the notion of an objective reality that we could know. The British empiricists, for instance, resisted the instability of an immanent frame by replacing a reality located in objective reason and created by a monotheistic all-powerful god with a reductive empiricism: all reality through reductive empirical methods can be objectively known through the empirical scientific method. In contrast, some Romantic scientists in Europe opted for a non-reductive, value-laden empiricism. From this "romantic empirical" position,[20] our very interiors, the lives of our minds, are tied up with the evolutionary exteriors (past and present) around us: our inner ecosystem is tied up with and bound to our outer ecosystem, and humans are tied up to other earthling's interiorities. The evolving assemblage of things does not allow for one to get out of this interior-exterior mix.[21] Each assemblage of life is the interiorization of exteriority, we might now say.

Objectivity, the view from nowhere, demands that we see the full assemblage all at once. It seeks to lay open the interior and the attempt at the exteriorization of that interiority turns the world inside out in an effort to understand it; little can survive such turning inside out. This is what extractive knowledge looks like: extracting value out of a thing, body, or phenomena so that it benefits the progress of (some) human beings with little or no concern for the given entity, nor its relations with other entities its entangled with.[22] This "turning inside out" can be seen (literally) in the dissection of animals and extraction of resources, or (figuratively) in the attempt to bring "dark masses" into the "light" through colonization. It can also be seen (intellectually) in the problem with mistaking one's own embodied cultural-historical-biological location as "objective:" our own subjectivity is projected as the reality. As argued in *Dialectic of Enlightenment*, this "turning inside out" or making all things in the image of monological logos/reason, destroys earth bodies and systems.[23] Claims of objectivity, as if one has an all-seeing eye that penetrates all things, often end in biotic death because we will never be able to turn things inside out (or make them in our own image) without killing them.

Most of Western epistemology was (and is) still caught up in the idea of moving out of the world of illusions found in Plato's cave and into the "real" world where the sun lays bare all illusions. Descartes' cogito, Kant's a priori ideas, the "thing in itself," and categorical imperatives, Hegel's synthesis between the material and ideal worlds, and even Heisenberg's "uncertainty principle" were all various ways of expressing this desire for reason and objective understandings to lead us to a view of the "real" world behind or under mere appearances. This all changes when humans are placed within an evolutionary framework. There is no "outside" that is real; what would an outside even look like? There are just different ways of understanding our entangled lives that provide the basis for different co-orderings of the worlds in which we live. Thus, Haeckel and many Romantics of his day understood nature is a part of culture and culture a part of nature. This not only had implications for thinking itself but also for social diversity.

One example of this new way of thinking about the world and the proliferation of possibilities for what life could become can be found in Haeckel's support of Magnus Hirschfeld's work in the late nineteenth and early twentieth centuries. Hirschfeld, a renowned biologist and sexologist of the nineteenth and early twentieth centuries, founded the Institute for Sexual research in Berlin in 1919. Before that time, Hirschfeld admired the work of and found support from Haeckel and other Romantic scientists, particularly in studies surrounding the evolution of forms in nature. They understood that once humans were placed into an evolutionary context, the idea of gender and sexual dimorphism and even heteronormativity would give way to a diverse number of ways in which human beings could be and become. Unlike Freud, Hirschfeld was not looking for the "heteronormative" reality behind hidden appearances, but rather at biology as an expression of different ways of being in the world. Forms are not superimposed but emergent from the process of nature naturing, and this is true of human forms as well. Human sexuality, then, would "naturally" be just as diverse as the forms of sex and sexuality found throughout the natural world. As one scholar notes,

> Haeckel's investigations in marine life had revealed that among the most primitive creatures hermaphrodism prevails, whereas creatures that have developed later tend to be diversified into males and females. Following Haeckel's recapitulation theory, Hirschfeld noted that this generalized observation about evolutionary or phylogenetic history is reflected by each individual animal's or human being's ontogenetic development, in which the embryo, at an early stage, is, 'ontogenetically bisexual' and only gradually develops male or female sexual characteristics. Since androgynous forms appear both phylogenetically and ontogenetically *before* sexually bifurcated forms, they ought to be recognized as *primordial*, or natural in the most original sense.[24]

The evolutionary context was essential for the Western mind to normalize diversity in sex and sexuality, whereas most cultures around the world before colonization by monotheistic cultures (really, Islam and Christianity) already recognized forms of gender/sex fluidity. This loosing of fixed forms from some transcendent reality enabled modern Westerns to rediscover gender/sex diversity as "natural."

Carl Linnaeus only a century earlier, for instance, advocated for species constancy and this is apparent in the way he transposed human relationships (from his own context) upon his description of the plant world. Plants were male and female, and had children, much like a modern bourgeois family.[25] I would argue that the lack of a strong form of evolutionary theory helped enable Linnaeus to impose European forms for thinking about sex and sexuality onto the world of plants. Of course, Haeckel, too, had his own cultural location that he imposed upon evolution in that he argued European humans, and particularly Germans, were at the top of the evolutionary tree.

The point here, is that Haeckel's deep observation of the diversity of sexuality in nature and his placing of humans into an evolutionary context meant that the understanding of God creating forms "in the beginning," including making humans as male or female (in the Priestly Genesis creation narrative), no longer made sense. In fact, gender dimorphism is not even evident in the garden narrative (the Yahwist narrative) of Genesis either, rather, the original Adam is of the mud, *Adamah*: a genderless mud creature. With an evolutionary understanding of nature the forms of organisms, including humans, evolve and emerge out of natural processes in which organisms act and react with other organisms. This is part of the reason Haeckel and others were so interested in embryology and morphology. He wanted to show the links and relationships between forms and order them all into a taxonomy.

ECOLOGY AND RELATIONALITY

Though Haeckel is often thought of as merely the popularizer of Darwin's theory, he built on the science with his own study of embryology, geology, and zoology. For Haeckel, the environment was quite important in terms of how a specific individual developed. He understood recapitulation as taking place across all species: ontogeny recapitulates phylogeny. This recapitulation theory is a kind of immanent version of hylomorphism, in which forms emerge from the entangled evolutionary process itself. His idea of recapitulation is that each individual organism, depending on where it is in the evolutionary tree, undergoes the developmental processes of all those species "below" (before) it on the tree. Differences between and within organisms emerge in the interactions with the ecosystems in which organisms find themselves. For Haeckel, then, changes in organisms are mostly a result of attuning to the surrounding environment. When these environments shift, different organisms will thrive. The plurality of human cultures could be explained by these environmental and geological differences as well. For Haeckel (among other monists and materialists at the time),

> dynamism was not limited to the individual, for the individual was embedded in the larger systems of its genealogical lineage, its species, and the collective of interacting species. Human society, too, was a dynamic system, moving along a developmental trajectory (*Entwicklungsbahn*). All living things were part of a great dynamic system. They were "linked together in the life of the great machine of the organic world, and this machine, too, seems to be advancing along a developmental trajectory."[26]

Haeckel's view, then, is somewhat in line with what Donna Haraway and others refer to now as sympoiesis rather than merely autopoiesis. *Sympoiesis* recognizes that individual organisms are a poor metric for understanding the variety of bodies and things in the world; rather organisms must be thought of in their coevolving lives together.[27] This is just as true at the atomic level as it is at the biological, chemical, and social levels. In this sense, Haeckel's understanding of "culture" *should* suggest that we look at the ways cultures evolve with one another and with the rest of the natural world. There is, we might say, an ecology of culture.

The emergence of forms and the interaction between organisms within nature, according to Haeckel, helped fuel changes in evolution, and for this reason Haeckel coined the term *ecology*. Though his theory of recapitulation itself is not in line with the "modern synthesis" between evolutionary theory and genetics, his basic idea of the entangled organisms influencing one another's development from within the ecosystems that they make up still makes sense. Our microbiomes for instance, suggest an internalization of the external ecosystem. From the very beginning, for Haeckel, the idea of ecology is linked with understanding diversity, relationality, and change (unfortunately for Haeckel, difference and change still had to be ranked). Human diversity and change were no different. Put into a contemporary context we might say the queerness of nature (or "ecology") has everything to do with the queerness of human forms as well: evolutionarily, biologically, and psychologically, forms interact and shift over time to such an extent that naming one life form this or that is a heuristic device that scientists use to understand the world in better or worse ways. We are, then, in continuity with, and not exceptions to, the rest of the natural world. For Romantic scientists such as Haeckel, nature was vital and alive. That is, all living things in nature have the ability to relate, and this everchanging relationality requires constant reattunement on the part of all entities involved.

Haeckel's own work, *Kristallseelen* (The Soul of Crystals), was an attempt to show that the patterns and "forms" of crystals are in continuity with the cellular and subcellular structures of biotic life.[28] Even if crystals were largely ordered "externally," their common patterns show both similarity and uniqueness of the kind that could also be used to explain patterns and forms

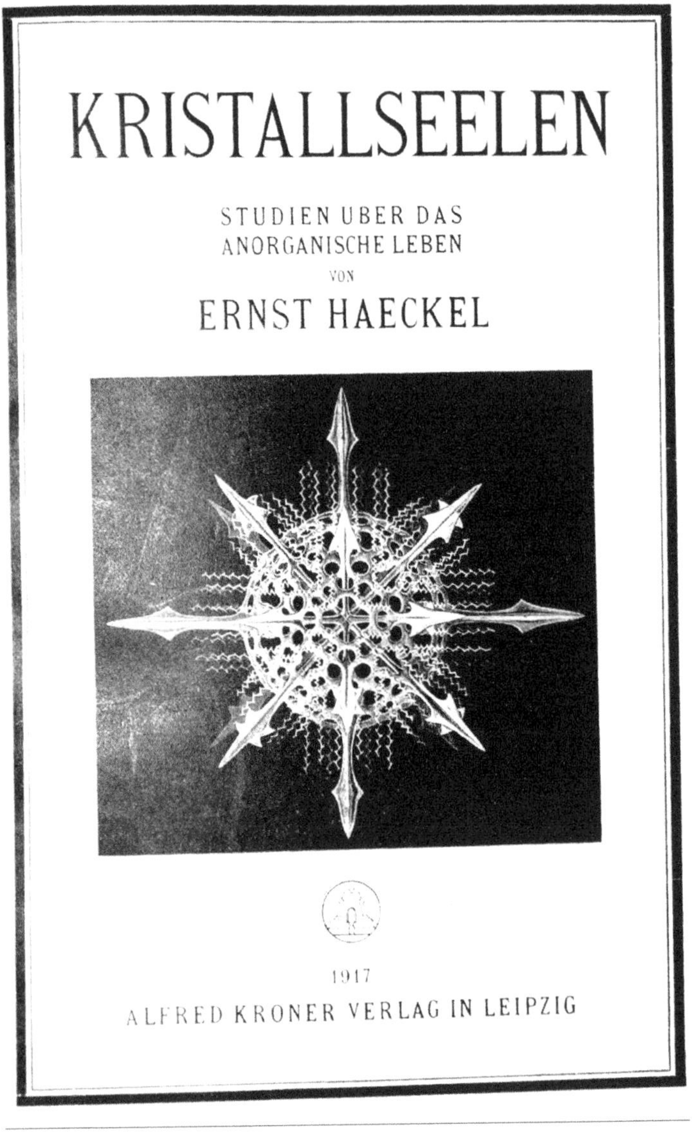

FIGURE 1.2 The cover of Ernst Haeckel's *Kristallseelen* (1917).

within biotic life. Furthermore, even at the level of crystals, all of nature in its interactions is very much "alive." As Olaf Breidbach notes, "The key idea [in *Kristallseelen*] was that the actual vital part [of a thing or organism] was not the individual organism, which after all was destined for death, but nature, which asserted itself in it."[29] In other words, it is through relations that all things have vitality, and nature naturing is the continuing process of this relationality.

Gustav Fechner, a contemporary and sometimes friend of Haeckel's, had a similar understanding of living nature as his *Seelenleben der Pflanzen* (*Soul Life of Plants*) suggested. In it, he argues for plant "interiority" and even "proto-communication," something contemporary botanists are confirming and understanding more and more. His point is, like Haeckel's, that if all things are to be explained immanently, then there are no external categories that keep us locked into place: no binary male-female; no hard boundary between plant, animal, and human; and no hard boundary between biotic and abiotic. He writes, for instance,

> Bei Mensch und Tier hängt an der charakteristischen Physiognomie, die ihnen zukommt, auch ein charakteristisch verschiedener innerer Bau, eine charakteristische Ordnung und Weise der Lebensprozesse. Eine verschiedene Seelenwirtschaft bedarf überhaupt einer verschiedenen Körperwirtschaft zum Ausdruck oder Träger, und der allgemeine Zug der Gestalt deutet nur äußerlich die eigentümlich zussamenhaltende und abschließende Einheit dieser inneren Wirtschaft dem Auge an. Und ganz so wie mit Mensch und Tier ist's auch mit der Pflanze.

> In humans and animals, the physical appearance and structure depends upon an inner structure and a characteristic order and life process. A different economy of the soul, needs a different

> physical economy to express or show the unique inner structure. And as with humans and animals, so it is with plants. (translation mine)[30]

Though there are clearly bad implications in terms of looking at the differences of things hierarchically and making conclusions, he's also placing humans on par with plants and animals, which is revealing in itself. In a way, Fechner is articulating a type of world-soul teleology, (everything is ensouled or alive), but one which may not be anchored in any external realm of universals and forms. Rather, as Terrance Deacon might argue today, these are emergent teloi in the ongoing process of nature naturing.[31] Fechner's work should be thought of in continuity with similar ideas espoused by scientists today in regards to emergence theory and plant communication, but with the critical insight that we don't need to rank evolutionary differences.[32]

Finally, Wilhelm Bölsche, a student, friend, and biographer of Haeckel, wrote about the evolution of love in nature. His purpose was both to ground human love in a wider understanding of nature—a type of precursor to E. O. Wilson's biophilia—and to show the complex lives of the rest of the animal world. As one commentator argues of Bölsche,

> The dramatic loss and then recovery of self, characteristic of the experience of love, *Bölsche* argues, is evidenced objectively at the very beginning of life, in the self-division of single-celled organisms, in the emergence of reproduction by sexual dimorphism, and ultimately in the totalizing aesthetic representation of the bond of all living things with all living things.[33]

Sentience, affect, love, consciousness, and agency, then, according to these Romantic scientists, are not limited to the human world, but rather pervade reality. The various levels of reality, we might say in today's parlance, are emergent and precisely not reducible. As is the case with the move from crystals to plants, and plants to animals, so is the case with the move from animals to humans. The intricate and intimate exchanges between organisms and different "levels" of life constitute ecology for Haeckel. Even the religious and scientific endeavors of humans are then emergent from the process of continuing evolution and part of ecologies.

RELIGION AND SCIENCE AS EMERGENT, RELATIONAL LEVELS OF ANALYSIS

There is, in this sense, an intelligence to life, but not of a traditional teleological variety. Rather, it is an emergent form of intelligence. Nature, for Haeckel and other Romantics, was about poiesis, or sympoiesis, and art. Humans do not merely impose ideas, forms and knowledge upon the world but are "fine-tuned" (or attuned) over millions of years of evolution to perceive the beauty of the natural world (not all in the same way, of course). In other words, we might say we are able to reattune to beauty in the world from within our various locations in the evolving process of life. According to Haeckel, religions (and philosophies) served an evolutionary purpose in as much as they helped human beings to become more aware of beauty, goodness and truth. However, he argued, the arrival of the natural sciences and particularly the arrival of evolutionary theory, which would put into context all sciences and humanities, meant that some old religions, monotheistic ones in particular, should be replaced by some form of what we might call religious naturalism, which he referred to as monism. He writes,

> Every great advance in the direction of profounder knowledge has meant a breaking away from the traditional dualism (or pluralism) and approach to monism. Ever more clearly we are compelled by reflection to recognize that God is not to be placed over against the material world as an external being, but must be placed as a "divine power" or "moving spirit" within the cosmos itself. Ever clearer does it become that all the wonderful phenomena of nature around us, just as the human body is only a small individual fraction of the great organized physical world.[34]

We must turn away from transcendent things and pay deep attention to the emergent patterns and forms in nature. Haeckel was certain that all peoples would benefit from close observation and study of the rest of the natural world from which we came, rather than reflecting merely on words, books, and ideas. It was for this reason that he, and the Monistbund in general, argued vehemently for education reform that would lead to more study of the natural world, and why he fought so vehemently against what he perceived to be the ignorance of theological and religious education. For Haeckel, promoting an evolutionary worldview was more than just a scientific necessity; it was also a religious or moral duty. The religiosity with which Haeckel pursued a monistic worldview in his arguments with dualists, idealists, and reductive materialists is evident in that his argumentation is often reminiscent of someone like Tertullian arguing against the Gnostics, or Augustine arguing against the pagans.

This argumentative style uncovers at least a twofold understanding of the nature of knowledge. First, knowledge must be fought and argued for. In other words, knowledge is always and already political, contextual, and shifting. It has implications for how we organize our lives politically, socially, economically and culturally. Though Haeckel refused to hitch monism to any specific political or economic platform, he did argue that the naturalistic worldview he was promoting took knowledge out of an elite knowledge-production system and placed it in the hands of a democratic endeavor. Everyone could "see" nature for themself, even if, Haeckel thought, there was a better and single method for doing so. Everyone could interpret data for themself, given the proper tools (though he would see European ways of doing so at the time as "better" than others). In my own language, everyone can reattune to the worlds around them through interpreting their own sensory inputs of that world. In this way, perhaps Haeckel would be a huge advocate of the "citizen science" that has become popular in the last few decades.[35]

There were no dogmas, thought Haeckel, that one must follow except the process of natural observation itself (and except for the dogmas he was unaware of stemming from his own unexamined location). Thus, knowledge production was not left up to a few clerics, academics, or statesmen, but rather was something to be experienced by all: Knowledge production should be much more egalitarian and democratic. This also meant that knowledge production should be a multiperspectival process: it takes people getting to know the rest of the natural world for themselves. Much like the Protestant Reformations encouraged literacy by translating the Bible into vernacular languages and allowing (more) people to read the texts for themselves, so knowledge of nature ought to be a collective process. This doesn't mean anything goes, but it does mean that knowledge is always co-constructed in one way or another.

If there is no objective, blank space to argue from, then it makes sense that we will inevitably smuggle in cultural, temporal, gender, racial, sexual, and all sorts of other biases/"locations" into our interpretations of the natural world, something Haeckel was not aware of. This is why we need critical discourses in our efforts to reattune to the worlds in which we live. We cannot nor should we try to escape our subjectivity, but we should try to recognize our embeddedness in an evolving community of diverse bodies, each one adding to the worlding process of which we are a part. This is the multiplicity, the pluriverse that emerges out of the process of evolution. Unfortunately, despite his philosophical resistance to certainty and bare facts, Haeckel thought he was somehow discovering a truer form of nature through his method, and that eventually MW science would get us to the whole truth about nature.[36] The temptation to mistake one's own location for objective space is great. What would it mean to refuse this type of objective space from which to see the world and truly embrace the interpretive framework in all our cultural, historical, and biological locatedness? What would it look like to critically reattune to the evolving worlds in which we live in a way that pays deep attention to the multiple perspectives of the becoming planetary community? I return to this in the final chapters of the book, but I think it at least means embracing some type of multiperspectivalism.

Haeckel went part of the way in arguing that including different perspectives on nature from the various emerging natural sciences gave us a better overall picture than any one science alone, and in placing humans and culture as a part of an evolutionary and ecological emerging process. However, he still held out hope that the MW sciences alone would eventually give us a more fine-tuned understanding of the natural world, and thus projected objectivity onto spaces and ideas that were co-constructed by nature-cultures and bio-histories. His progressive understanding of time led to a type of delayed objectivity that he could at times glimpse and grab hold of: a messianic "to come" understanding of objectivity, but one that would come with certainty.

One cannot fully separate cultural location from understandings of nature, even if it is a future projection of the arrival of an objective view. This is not in itself a bad thing: knowledge is just contextual. When this is not recognized or acknowledged, however, it allows biases and locations to be projected as objective truths. At times contemporary religious naturalists, environmentalists, and other self-proclaimed progressives read and sound as if they have arrived at the real truth and everyone else just needs to catch up. I think we can learn from Haeckel's mistakes in this regard, as a cautionary tale for anyone who seeks to develop an immanent worldview based upon the evolving natural world of which we are a part.

HERMENEUTICS OF NATURE: SCIENCE AS A TOOL FOR REATTUNING

From the "clear and present" present, we cannot see the layered worlds in which we live all at once. The layers only become clear in retrospect, which is why the worlds of those whom we look back into history upon seem "all mixed up," or at least outdated from our clear and present positions. The same is true when looking at Ernst Haeckel and other Romantics of the nineteenth century. What we might consider now as a mixture of Romantic nature spirituality, science, and "pseudoscience" (I dislike that term), seemed to them to be the integration of a clear and present life. In looking back, then, we cannot make the mistake of the primacy of the present. This book attempts to sort through some of the layers of the construction of what has today become a naturalistic view of the world. I have tried thus far to look at what evolution and ecology have meant for hermeneutics and knowledge, but here I want to end this chapter with what this might mean for *doing* science. Again, as elsewhere, Haeckel seems all mixed up from where we now stand: at one moment he states that there are no bare facts while at the next he is declaring that *doing* science is merely looking at the natural world and allowing it to reveal itself to us.

From where Haeckel stood, this made sense. Though he understood that evolution placed humans within nature, and that meant culture and ultimately science itself were parts of nature naturing, he didn't yet have the tools to critique his own socio-historical location. Part of this was due to the importation of progressive and cumulative understanding of time that he took from the background monotheism of his culture. In addition, there had not yet been huge civil rights movements, much less postcolonial analyses of life: the evolutionary theory that was opening up the world for a plurality of perspectives was still emergent and theorized from a colonial, heteropatriarchal mindset. Finally, the world had only just begun to be radically intertwined on a daily basis through faster communication, transportation, and production technologies, not to mention the first of two world wars late in Haeckel's life. What I am suggesting is that it took a lot more then than it does now to recognize the racial, cultural, and other types of embodied locatedness of different ideas. Yes, there were other scientists who strongly disagreed with his racist hierarchy of cultures in interpreting the evolution of humans.[37] But there were also many who hated the idea that humans emerge out of the same stuff and have some sort of kinship to all other animals. German Romanticism and Darwinian thought came together, for Haeckel, to help challenge this type of human exceptionalism. Yet, these challenges did not extend to his thinking about non-Western, nonwhite cultures. His racist, ablest, and antisemitic comments reveal his own understanding of time as linear and progressive. From this perspective, he understood some humans as "more progressed" than others.

None of this analysis is meant to excuse Haeckel; it is meant, rather, to help explore why some critiques of oppressive ideas don't necessarily lead to an intersectional analysis: Why are some women racist in sexist societies? Why are some gay men misogynists or racist in heteronormative ones? Why do some Black men have heterosexist or sexist attitudes? Why does anyone who is marginalized not automatically empathize with all the life that is subjugated by anthropocentrism? These are thorny questions, and intersectional, historical analyses help get at some answers to these questions. This underscores the need for the "critical" part of a CPR. The avoidance of a critical, intersectional analysis might have something to do with that ever-elusive (subconscious?) desire to escape hermeneutics and gain access to an objective, real, and unmediated reality. The road to progress in knowledge has many casualties that are ignored and buried. The lure to take a rest on the comfortable bed of objectivity is a strong one. Anytime we give in to this, from wherever we are embodied, we make the mistake of smuggling in cultural, historical, embodied biases and projecting them as objectivity. This is precisely why we need critical theories in the process of reattuning to the worlds around us. There is no direct, unmediated experience of reality or ground of being, or whatever we want to call it: rather every entity, person, or thing that exists is involved in co-creating the shifting grounds of what is. Mistaking one's experience as unmediated has devastating consequences because it distorts this pluriform, shape-shifting planet into the image of one part. Without critical theories, an immanent romanticism can be put to the use of one ideology or another: all other relations are then related to a central ideology, person, place or thing. A CPR must then include the multiple perspectives and worlds within the entire planet, and some understanding of how different bodies experience those worlds and the planet differently.

Olaf Breidbach has argued that Haeckel ingested Goethe's understanding of a progressive evolution but unlike Goethe's, his was an open and emergent understanding of evolution. Breidbach argues, that Haeckel was "putting forward a typology, but, unlike Goethe, not revealing a predetermined structure of nature but only tracing the phases of an ultimately open process in its classification of types."[38] Haeckel had gotten rid of the idea of eternal forms, in other words, but held on to the idea that there is a qualitative directionality in evolution. This, I argue, still provides a place for values and beliefs that seem to be objectively better or more advanced but are, in actuality, culturally and historically located. What happens when we begin to think about a hermeneutics of nature without resting spots; without the smooth, eternal, unchanging space above the fray, and without certainty in the progressive direction of life? What happens if the evolution of life is a pluriform reality whose parts are constantly reattuning to one another? I don't think the answer is complete relativism but rather an open, emergent common grounds, which acknowledges the undercommons that make any given grounds possible. Again, critical theories are so important for uncovering the various perspectives and embodiments that make up the common grounds of any given worlding.[39] This is what I insist on the *critical* part of a planetary romanticism.

From this perspective, objectivity and universality are redefined as multiperspectivalism. As Catherine Keller suggests, following Nicolas of Cusa (among many others): more voices and more perspectives leads to more knowledge of our common, uncommon, and undercommon grounds. It is not a unity but a multitude that makes up these grounds. She writes, "What if we go lower yet, get common enough for an uncommon solidarity with the undercommons? Beneath foundations, where the intensity of my issue rocks into the movement of a multitude, issuing not in unity but in an immense coalescence?"[40] At the base of all things is not a single unity but plurality; in the case of Haeckel, this was a triune substance: matter, energy, and sensing. This trinity may be too small still for our contemporary globalized world, but it functions as an alternative historical example to that of reductionism and universalism. A plurality rather than mere unity or oneness (despite his calling it "Monism").

Once history and consciousness are placed into an evolutionary context, Haeckel understood there were no bare facts. But he thought certain humans—and certain methods, viz., Western cultures and science—had been highly fine-tuned through evolution to see nature as it was really unfolding. The entire evolutionary trajectory, it seems, led right up to Haeckel's perspective as the most advanced one. He did not understand that his German, white, European, Christian–influenced perspective of science was but one interpretation of the natural world. As such, it should be understood as just that: an interpretation among many that can be placed into dialogue. A new type of pantheism or animism that could dialogue with other relational, animistic, and pantheistic ideologies and narratives.

This emergent and multiperspectival understanding of "nature naturing" doesn't mean we can't agree on things—we all need oxygen to survive, we are all earthlings, our earth is within a vast universe (maybe even a multiverse)—but it does mean that there is no "above the fray" and that more perspectives give us a better understanding of things/events in the world. Thus, the old false dichotomy between objectivity and relativity is revealed as false precisely when we begin to think of our thoughts/knowledge/ideas as part of evolutionary and ecological contexts, as a thinking in place.[41] Or, as Walter Mignolo suggests, we move from the colonizing "I think therefore I am" to the decolonizing "I am where I do and think."[42] By placing humans in an evolutionary perspective, science not only makes culture and religion radically immanent, but also science itself. This means hermeneutics are important "all the way down," and "all the way out."

2

THEOLOGY, CRITICAL ROMANTICISM, AND THE CONSTRUCTION OF A NONREDUCTIVE NATURALISM

In dem universalen Begriffe der Substanz—als des ‚wirklichen Weltwessens'—vereinigt unser naturalistischer Monismus drei untrennbare Attribute oder Grundeigenschaften: die raumerfüllende Materie (=Stoff), die wirkende Energie (=Kraft) und die empfindende Weltseele (=Psychom).

In naturalistic monism, three inseparable attributes are understood as basic properties of substance or real knowledge of the world: space-filling matter (stuff), active energy (force), and the sensing world soul (psychom).

—Ernst Haeckel, *Gott-Natur (Theophysis), mit Kommenter Nachdruck von Olaf Breidbach und Uwe Hoßfeld* (translation mine)

Ernst Haeckel was an early, modern adopter of a full-blown naturalistic worldview as described in the epigraph.[1] He was not only concerned, in other words, with studying nature per se but, like Alexander von Humboldt before him, with understanding how all things human fit within the emerging idea of "nature" during the nineteenth century. In doing so, he was bound to stumble upon some of the largest arguments of the "culture wars" during his time, which included questions of whether or not humans and other species were somehow created in the beginning, and whether or not religious understandings of nature should be supplemented, if not supplanted, by an evolutionary framework in which all life is interrelated and changing. Evolution, in this sense, becomes a kind of religion: one which has much more in common with animisms, pantheisms, and relational ontologies of Eastern traditions, and one which "corrects" the problems of Western monotheisms as Haeckel understood them.[2]

In this chapter, I explore how religious ideas, in particular Christianity, influenced the way in which he viewed nature. Haeckel's understanding of "nature" was also influenced by German Romanticism, along with the Eastern traditions that influenced Romanticism, including Islamic thought, and the pantheism that fueled his own monistic philosophy. Scholars such as Kate Rigby and Kocku von Stuckrad have argued that Romanticism and its corresponding pantheism is perhaps the leftover remnants of what we might call European Indigenous traditions.[3] In this sense, evolutionary theory is a "return" (though not purely) to such Indigenous roots. Something like this type of pantheism or animism fits better with Haeckel's own understanding of the natural world. But Haeckel was also deeply shaped by theologically filled understandings of "creation" that were smuggled into his understanding of "nature." These theological imports included a "progressive" understanding of evolution and time that he read into the evolutionary record, fueled by a desire that certainty would be attained by science and the great mysteries or riddles of the world would all be answered. In other words, we would eventually arrive at universal truth (a latent form of teleology or even eschatology). It seems the more he tried to critique theology, the more he created a mirrorlike image of monotheism in his version of religious naturalism.

This chapter, using Haeckel as an example, explores the ways in which science and religion are always and already together when developing a worldview, in this case a naturalistic one that includes elements of theology and elements of Romanticism (and the religious influences therein). In a critical planetary romanticism both science and religion (always already mixed) are as good as they help to us reattune to the worlds around us. Studying the religious influences on Haeckel's work is relevant because it provides an example of how ideas materialize in the world around us.[4] Haeckel's monistic ideas had a politic, an ethic, and a method for educating the public that still have rippling influences upon us to date. To argue these points, I first start with a brief lineage of romantic thought from Spinoza to Haeckel, which constitutes a major philosophical, theological, and scientific tributary to the development of his triune monism. It is important that Haeckel draws from one of the nondominant traditions within so-called Western thought that is found in Heraclitus, Nicolas of Cusa, Bruno, Spinoza, Herder, and Schelling viz. pantheism. Second, I turn to the ways in which Schleiermacher and Kant helped make a shift toward the "turn to experience," which also was a main tributary for Haeckel's understanding of the world, especially in terms of loosening authority away from tradition (and mainly theological traditions) and toward investigation of experience and the senses (a process of deterritorialization). Third and finally, I look at Haeckel's triune monism more closely in the context of the Industrial Revolution and its ills, and the Romantic reaction to that. This context, and the growing reductionist/productionist and progressive model of understanding nature combined with the insufficient explanatory power of theological dualism, encouraged in Haeckel's thought a third way that we might think of today as a precursor to emergence theory and nonreductive religious naturalisms.

ROMANTICISM FROM SPINOZA TO HAECKEL

> *In endeavoring to steer a path between the Scylla of materialist reductionism and the Charybdis of idealist hubris, Schelling developed a genetic theory of subjectivity, according to which human consciousness emerges out of organic nature, just as organic nature emerges out of inorganic matter.*
>
> —Kate Rigby, *Topographies of the Sacred: The Poetics of Place in European Romanticism*

The tone of Ernst Haeckel's writings, especially his very popular *Riddle of the Universe,* must be set between two primary bookends, similar to the bookends of Schelling in the quote above.[5] The first is his own self-placement within a stream of reformist and Romantic thought starting with the likes of Bruno, Galileo, Copernicus, Martin Luther, and Newton, on the one hand, and Spinoza, Alexander von Humboldt, Goethe, Herder, and Schelling, on the other. In other words, he situates himself, and the whole project of understanding the natural world within the emerging natural sciences of the nineteenth century, in line with the continuing process of reform and enlightenment begun by the revolutions in science, and the reformations begun by Luther (et al.), and in line with what would become known as German Romanticism. He writes, for instance:

> Naturally, through the long night of the Dark Ages it was rarely that a brave free-thinker ventured to express an opinion to the contrary: the examples of Galileo, Giordano Bruno, and other independent philosophers, effectually destroyed all freedom of utterance. Heresy only became possible when the Reformation and the Renaissance had broken the power of the papacy. . . .
>
> It was the gigantic progress of biology in the present century, and especially in the latter half of the century, that finally destroyed the myth. The establishment of the theory of descent and the cellular theory, the astounding discoveries of ontogeny and experimental physiology—above all, the marvellous progress of the microscopic anatomy of the brain, gradually deprived athanativsm of every basis; now, indeed, it is rarely that an informed and honorable biologist is found to defend the immortality of the soul.[6]

This self-understanding that recapitulates the myth of the "Dark Ages," erasing the influences that Islamic thought and scholarship had for bringing about what we commonly refer to as the Scientific Revolution (a necessary myth for understanding his own beliefs as "progressive"), is quite telling. It places him in his own context as a revolutionary thinker fighting against what he perceived to be the evil dogmatic slumber of traditional Christian (and especially Catholic) theology. In a sense he was carrying on the tradition of the Reformation through his scientific method, and he clearly thought other Romantics were doing the same. He makes repeated, hyperbolic anti-theological (and especially anticlerical) statements in order to make his point that the new methods emerging in the sciences are far better at producing knowledge of the world than any theological method and can provide the basis for a religion: monism. He writes, for example, "No cosmic problem is solved, or even advanced, by the cerebral function we call emotion. And the same must be said of the so-called 'revelation' and of the 'truths of faith' which it is supposed to communicate; they are based entirely on a deception, consciously or unconsciously."[7]

Dualistic Christian theology, however, was not the only form of thinking that received his ire, but rather all forms of idealism. He admired Luther for the role he played in undermining Catholic dogma and for the role he played in the turn toward individual reason. For him, Luther's mistake was that he held on to the Bible rather than hitching his wagon to sensory experience of the natural world, and reforming religion based upon that experience.[8] Likewise, he greatly admired Kant but thought Kant's mistake was that he gave experience authority with one hand and took it away with the other hand when it came to understandings of God and theology, which Kant argued were beyond our capacity for understanding.[9] He did follow Hegel's understanding of the progressive evolution of ideas and history because it matched up well with Darwin's thought, but he thought Hegel's mistake was that he began and ended with idealism, rather than with the material world.[10] In fact, people would charge Haeckel with promoting Marxist socialism with his version of scientific monism, despite Engel's dismissal of Haeckel, and Haeckel's arguments that his monism does not support any particular ideology.[11] As I argue throughout this book, the (false) assertion that Haeckel's evolutionary monism supports Marxism would later be used by early creationists to make the facetious connection between evolution, Marxism, atheism, and Nazism, on the one hand, and creationism, capitalism, theism, and democracy, on the other. Haeckel in fact saw any economics or politics—idealist or materialist—tied to the experimental method as a form of imposition that could obscure the method and as thereby erroneous.[12] His iconoclastic attitude toward materialism, idealism, and dualism proved to be somewhat of a Rorschach test for other scholars of his era. In the end he was charged by philosophers and theologians with reducing everything to materialism, and he was charged by scientists with interjecting religion, or at least some type of vitalism, into science.

Haeckel felt most at home with the likes of pantheistic thought and German Romanticism. Of all the thinkers he drew from, perhaps the most important were Spinoza, Alexander von Humboldt, and Goethe.[13] These thinkers were, of course, influenced by mysticisms and relational ontologies of Eastern traditions, particularly Vedic ones. Spinoza was influenced, for example, by Kabballah, while Goethe, Schlegel, and von Humboldt were all influenced by the influx of Indic and Islamic literatures, even if in an Orientalist sort of way. There would be no "Scientific Revolution" nor a "European Romanticism" without European philosophers and scientists reading texts from the Arab and Indic worlds.[14] In many ways, Romanticism, like the natural sciences, is from its inception a planetary way of thinking, drawing from Indigenous, Buddhist, Muslim, Hindu, Jewish, and Christian traditions.

Goethe and Spinoza, both of whom benefited from the aforementioned multicultural resources, seem to be two of the most important sources for Haeckel in thinking romantically about nature. From Spinoza, Haeckel took the primary idea that one can talk about the whole of life from within a naturalistic framework, namely that provided by the theory of evolution.[15] Of course, Herder and others helped develop this type of *Naturphilosophie* along with Shelling, Goethe and many others. They all believed in the vitality and the unity of nature, and that humans were a part of it.[16] Spinoza's pantheism, falling within this same tradition, is not without its multiple interpretations.[17] Though some argue his pantheism was closed, deterministic, and thus brought all the problems that come along with wholism, others argue it was more open and falls in line with the open pantheism that flows from Herder to Goethe and Shelling. In this sense, as Robert Richards notes, "there is no matter without mind [Geist] (that is, without that necessary power determining matter), nor any mind without matter. Rather there is neither mind nor matter in the usual sense. There is only one thing that is both simultaneously. To accuse this opinion, which rests on observation, of materialism is as perverse as charging it with spiritualism."[18] Whatever the interpretation, Spinoza's pantheism provided Haeckel with a historical example of what a genuine naturalistic worldview, including discussions of philosophical and religious ideas, might look like. He writes,

> The form of monism which I take to be the most complete expression of the general truth, and which I have advocated in my writings for thirty-eight years, is now generally called hylozoism. This expresses the fact that all substance has two fundamental attributes; as matter (hyle) it occupies space, and as force or energy it is endowed with sensation (cf. chapter xix.). Spinoza, who gave the most perfect expression to this idea in his "philosophy of identity," and most clearly treated the notion of substance (as the all-embracing essence of the world), clothes it with two general attributes—extension and thought.[19]

In light of the cultural battle of Haeckel's time then, the choice of something like Spinoza's God of Nature over that of the church's God, was quite simple. Haeckel (though philosophical) was not a trained philosopher and did not get into deep explications of Spinoza's work, and perhaps would have disagreed with the more reductive interpretations of it. Regardless, Spinoza's shift to an immanent framework was key to Haeckel developing a viable naturalistic worldview. Part and parcel to working from within such a naturalistic worldview was, of course, the direct study of nature. Toward this end, Haeckel found intellectual heroes in the German Romantic scientists, and particularly in the person of Alexander von Humboldt.

More than any other intellectual pioneer before him, von Humboldt had really constructed a contemporary idea of "nature," that was interdisciplinary (bringing together sciences from multiple fields) and observational, in that his knowledge was largely due to the experiences of his extensive periods of time exploring various natures, places, and peoples around the planet, and cataloging the different life forms in detail. Andrea Wulf's book *The Invention of Nature* provides a thorough summary of von Humboldt's extensive travel and his efforts to cobble together a common view of "nature."[20] Haeckel, in a sense, saw himself as extending Humboldt's work through the theory of evolution, and particularly Darwin's version of it. Haeckel's detailed studies of various species, life forms, and even abiotic patterns of nature found in many of his books, would go on to influence the emergence of art nouveau in the late nineteenth and early twentieth centuries, part of what I think was an aesthetic reattunement to the new nature described by evolutionary theory.[21]

Sadly, Haeckel's various cultural evolutionary trees and rankings, based largely on the emerging views of progressive philological studies during his time, which came out of emerging Indo-Germanic studies, provided tragic support for the rise of cultural and scientific ethnocentrism in the lead up to the two

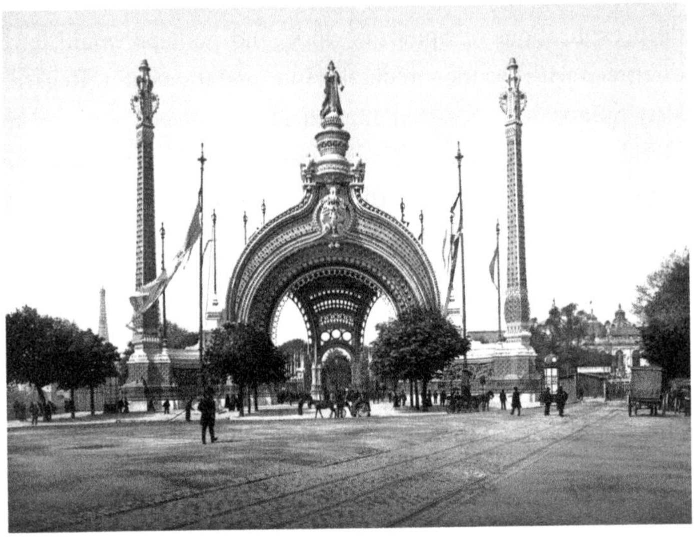

FIGURE 2.1 Porte Monumentale, Paris, 1900. The gate was influenced by Haeckel's *Artforms in Nature*. (Courtesy of Wikimedia Commons)

world wars of the early twentieth century. Though his was not a racism based upon anything like biological race, and he had no ideas of purifying populations nor eliminating people based on common attributes, he nonetheless saw more or less progressed humans emerging from two different species, "the woolly haired" and the "straight haired."[22] This will be discussed further in chapter 5. Here I emphasize the emerging consensus of the connections within nature that people such as Spinoza and von Humboldt made possible, and how this influenced Haeckel's understanding of human beings as a part of nature.

It was the deep study of various landscapes and organisms, of being "out" in nature, that helped von Humboldt and others of his ilk to begin questioning received traditions about nature and the role of human beings within nature. The sheer diversity of nature from what is now the Indonesian archipelago to what is now the South American continent, to the hills of Jena, made it quite clear that nature was far more diverse and complex than previous generations had imagined. At the same time, Haeckel argued that the divisions between chemistry, biology, geology, and zoology were such that too many scientists got stuck within their own methodologies at the expense of drawing connections. This connective type of knowledge, the same that von Humboldt used to help construct the more romantic Western idea of nature, was important for Haeckel's understanding of nature and the place of humans within it.[23] For instance, Haeckel argues that the direct study of nature from many different disciplines is better than the training of people for a specific profession and from within a single disciplinary background. He writes:

> The chief aim of higher education up to the present time, in most countries, has been a preparation for the subsequent profession, and the acquisition of a certain amount of information and direction for civic duties. The school of the twentieth-century will have for its main object the formation of independent thought, the clear understanding of the knowledge acquired, and an insight into the natural connection of phenomena. If the modern state gives every citizen a vote, it should also give him the means of developing his reason by a proper education, in order to make a rational use of his vote for the commonweal.[24]

Knowledge, for Haeckel, was political. The separative and reductive understanding of the natural world was detrimental to cultivating a naturalistic worldview, and was thus also detrimental to civic life, according to Haeckel. He would likely shudder at the fragmentation of the university today: both within the sciences and between the sciences and the humanities. At the same time he would likely applaud interdisciplinary approaches to understanding the world around us, such as religion and science, religion and nature, the environmental humanities, and posthumanism(s). What was needed, in Haeckel's mind, was a re-collecting of ideas and a rereading of nature (a form of *re-legare/re-ligare*). In other words, he was reattuning to the world around him in light of evolutionary theory and not just in a rational way, but also in emotional, aesthetic, ethical, and pedagogical ways. In this sense, his monism was indeed a religious naturalist's attempt to make sense of the world by bringing ideas together into a coherent meaningful way.

For Haeckel, von Humboldt was a pioneer in bringing ideas from various sciences together into a coherent understanding of "nature."[25] Though von Humboldt rarely ventured into religious and philosophical ideas, he did note the connections between people and nature. Everything for von Humboldt, as for Haeckel, was interrelated and connected. Von Humboldt's way of expressing this was through his *Naturgemälde* (nature paintings). As a painter, he was like Goethe before him, interested in the way bodies were affectively influenced by their environs. What von Humboldt started, Haeckel sought to expand on with zeal: he wanted to place all things human within this naturalistic framework, and therefore include the natural sciences, psychology, and even philosophy, theology/religion, and art. Humans were then a part of the nature painting, not outside of it. As Andreea Wulf argues, Humboldt, following Goethe, interpreted "the natural world as a unified whole that is animated by interactive forces. This new way of thinking changed his approach. If everything was connected, then it was important to examine the differences and similarities without ever losing sight of the whole. Comparison became Humboldt's primary means of understanding nature, not abstract mathematics or numbers."[26]

It was Goethe above all others who influenced Haeckel in terms of a view of nature that might include human bodies, affects, and aesthetics. He developed a "tender empiricism" that was a nonreductive reading of the rest of the natural world. Goethe's "zarte Empirie" or "tender empiricism" (what Robert Richards has called "Romantic biology")[27] suggests an "empiricism that intimately merges with its object and through this very identification becomes proper theory."[28] Goethe caught in literature the changing ideas of humans from that of an exceptional, almost divine animal within the rest of the natural world toward that of humans as a part of the rest of the natural world that was alive and animated. As Richards notes, "What Darwinism offered instead of traditional orthodoxy, Haeckel contented, was Goethe's religion: a monistic religion of humanity grounded in pantheism."[29] Embodiment, emotions, desire and eros became important themes of Goethe's literary and scientific writing. The descriptions of the natural world in Goethe's work were not just mere backdrops for stories to unfold, but part and parcel to the plot and purpose of what he was trying to say. In a similar sense, as James Miller's argues in *China's Green Religion* from a totally different context, Goethe was shifting to an understanding of nature that suggests nature was not just something we get stuff from and use, but rather it is what makes humans human.[30] It is that which makes our own human forms and expressions possible, rather than those human forms and expressions being understood as that which makes humans exceptional to the rest of the natural world.

Haeckel built upon Goethe's and other Romantic's immanent thinking and extended it to religions, cultures, languages, and all things human in general. It was for this reason that he took on what he thought to be misleading idealistic philosophies, and misleading dualistic, theological thinking. Humans were, through and through, a part of the rest of the natural world and this meant that our art, our cultures, our language and our religions were also a part of nature. In different ways, Wilehlm Bölsche, Gustav Fechner, and Ernst Haeckel, all of whom I would call Romantic scientists, or "romantic empiricists,"[31] were trying to bring ideas and other things immaterial (aesthetics, forms, morals, values) to an immanent frame as part of an overall naturalistic worldview, and this would have been impossible within their cultural-historical location without the help of earlier Romantics such as von Humboldt and Goethe.[32] As von Humboldt once wrote to Goethe, "'Nature must be experienced through feeling,' [and] . . . those who wanted to describe the world by simply classifying plants, animals and rocks 'will never get close to it.'"[33]

As Haeckel was developing his romantically influenced triune monism, Bölsche took seriously the more eco-erotic strands of Goethe's (and Herder's) thought and wrote a book about his observations of the evolution of love in nature. He traces "love" as an attraction that is present throughout all life, even in the single cell, and even discusses some of the sex lives of various organisms (including the "love darts" of snails). He writes, for instance, "the tremendous romance of love-life in Nature, with all its madness, its pain, its blindness and brutality, and all its sweetness and all its light, is the pilgrimage of an ideal."[34] Bringing emotions, ideals, energy and matter together into a concept of pilgrimage, Bölsche hints at what Herder and Goethe before him were seeking: a type of romantic empiricism. As Dalia Nassar explains, for these early Romantic empiricists "matter could not be merely passive (moving according to the laws of motion), or blind (acting without any direction) but active and capable of *internal* transformation."[35] The whole of life, then, might be considered an evolutionary pilgrimage that includes the emergence of humans and all things human.

In a sense, just as Haeckel was returning thought to the natural world, Bölsche was tracing love, and other affects, in the rest of the natural world as emergent features central to the evolution of life rather than as something foreign and human, imposed upon the material world. Similarly, Gustav Fechner was arguing for the vitality of the plant kingdom in his *Nanna oder über das Seelenleben der pflanzen* (Nanna: About the Inner Life of Plants). Indeed, he argues, like Schelling before him, that there is something like a world soul,[36] and it is this world soul that coproduces, as part of the physical world, the realities that we exist in and experience.[37] Again, this is an earlier attempt to do what the so-called new materialists are attempting to do today: return vitality, agency, and vibrancy to the rest of the natural world, after the "Western modern mind" tried and failed to separate humans from nature. As Richards notes, "This organic conception of nature—given currency by Herder, Goethe and Schelling—opposed the mechanical ideal stemming from Descartes and Newton. Within the tradition of *Naturphilosophie*, nature ceased to be mere product of the creator's designs but itself became producer—of itself."[38]

Above all else, what the Romantics (scientists, artists, writers, dancers) pointed to, along with the emerging evolutionary framework, was that we know what we know through our creaturely embodiment and not through access to some sort of ethereal forms of knowledge as previous philosophers and theologians might have imagined. This is one reason, again, that Haeckel influenced artistic movements like art nouveau and modern dance (especially through Isadora Duncan).[39] In developing his triune monism, Haeckel's disagreements with Kant, even if based on an erroneous interpretation, and the influence of Friedreich Schleiermacher's turn to experience in theology (among other things) became important conversation partners.

THE TURN TO EXPERIENCE IN PHILOSOPHY AND THEOLOGY

In many ways the "collapse into immanence" away from transcendent sources for ideas, origins, and values, afforded by the placing of human's into an evolutionary framework, was a precursor to what Haraway has called "tentacular" thinking: a collapse into the earth community by which the only way to understand ourselves is to think along with the multiple earth others that make our own embodiments possible. This is to think beyond the abstract, exceptional human and human history and think with the rest of the evolving planetary community. Haraway writes, "History must give way to geostories, to Gaia stories, to symchthonic stories; terrans do webbed, braided, and tentacular living and dying in sympoietic multispecies string figures; they do not do History."[40] As embodied creatures, we have to account for all of our experiences with others in the world, both seen and unseen, and not just of the human and the linear event-built-upon-event understanding of human history. This is what it means to let go of the transcendent "no space" of Western-style Enlightenment thinking: there is not one Archimedean point at which we all arrive and agree on, or from which we can stand and force all of history into a tunnel of time that leads up to that Archimedean point. Rather, from an immanent, evolutionary understanding of human beings and the planet, there are multiple perspectives and multiple stories. The more perspectives and stories we have of a given world or terrain, the better understanding we have of that world or terrain. Haeckel and other Romantic scientists of his day likely wouldn't have gone this far in terms of pluralism, but they took seriously the notion that humans are interrelated with all other life on the planet not just bodily, but also subjectively, agentially, aesthetically, and etiologically. In other words, whatever it means to be human must be a result of our relationship with the rest of the natural world. As such, we must think with rather than merely for or about nature. This type of thinking requires constant reattunements to the evolving planetary community and is a precursor to something like the postmodern condition and to something like the new materialisms.

Part of this so-called collapse into immanence also involves a critique of strict disciplinary boundaries. Strict disciplines require a no-space from which one can stand and separate the world out into distinct things and ideas. If everything is entangled, then such sharp boundaries and distinctions do not exist in the world. The boundaries around certain fields, disciplines, or entities are much more ecotonal and porous. Haeckel, like von Humboldt, Goethe, Darwin, Marx, Freud, and William James (among many others) crossed multiple disciplinary boundaries in searching for explanations for experiences of the world. We can, from where we stand, identify Haeckel the philosopher/theologian, Haeckel the scientist, and Haeckel the artist, but these would have been all mixed up together in his own person, along with any contradictions they may have produced. This was indeed the time in which the sciences were becoming solidified into a single tapestry, and the fights over whether *Naturwissenschaften* or *Geisteswissenschaften* would give us ultimate, Enlightenment-style objective truth were at full pitch. Still, these things were thought together, something which is hard to understand from contemporary Euro-Western, hyperdisciplined contexts. It's no surprise then that Haeckel the scientist would engage with many of the best philosophical and theological minds available at his time.

One of Haeckel's primary interlocutors in developing a proper, nonreductive but materialistic epistemology was Immanuel Kant. Kant had, after all, reflected on the process of "globalization" and on "cosmology" in ways that were ringing true during Haeckel's era.[41] Furthermore, he was the primary philosophical touchstone of the nineteenth and early twentieth centuries. Everyone had to go through Kant to get anywhere philosophically, and Haeckel as a philosopher of the emerging naturalistic worldview was no exception. He agreed with much of Kant but was critical of Kant's idea of the *ding an sich* and a priori ideas. As Kate Rigby notes,

> In declaring things in themselves (*Ding an Sich*) out of bounds to human knowledge, Kant had dealt a valuable blow to a certain epistemological hubris: the assumption, that is, that philosophical reflection or scientific investigation would be able to reveal nature in the raw, unmediated by human perceptions and projections. However, in identifying reason as transcendental or *a priori*, that is, not derived from experience but rather constitutive of experience, Kant also effectively severed it from the phenomena in the natural world that it posited as its object of knowledge.[42]

Haeckel's disagreements with Kant all stemmed from the idea that placing humans within an evolutionary framework meant that all knowledge is a posteriori knowledge and that there could be no "ding an sich" in the way that he interpreted Kant to mean it. He writes, for instance, "What seems to us today to be an innate capacity, or an *a priori* quality, of our phronema, is really a phylogenetic result of a long series of brain adaptations, formed by *a posteriori* sense perceptions and experiences."[43] Not only does this type of thinking serve as a precursor to Foucauldian-style genealogical analysis and postmodern turns toward hermeneutics, it also means that one can only learn through their embodied experience of the world. This evolving embodied experience of the world also meant there could be no "ding an sich" understood as a separate entity.

Whether or not Haeckel was correctly reading Kant's "ding an sich" as an essence beyond which any pure reason might be able to go, that is how he understood it. For Haeckel, such an essence meant that there must be something transcendent from the process of nature naturing. In other words, he understood it to mean there must be a soul or a cogito that is beyond the rest of the natural world. His rejection of this type of essence (as he perceived it) does not mean that there is no mystery or wonder in the world; rather, because the world and all entities therein are constantly changing, emerging toward an open future, no one embodied manifestation of the flow of life can be cut off from this process and fully known in itself. Though Haeckel was confident about solving the riddles of nature, he was also aware that individual organisms and entities could never fully be known by the gaze of human and natural sciences, and that the process of studying nature naturing provoked wonder. The study of nature, then, à la Schelling, was not the discovery that humans are the consciousness of nature, but rather that we are part of a nature that is alive, awe-provoking, and mysterious. As Schelling writes and I think Haeckel would agree, "For the consciousness of man [*sic*] is not the consciousness of nature . . . Far from man and his activity making the world comprehensible, man himself is that which is most incomprehensible."[44]

This embodied experience, because it was tied up with an evolutionary history of other organisms, meant also for Haeckel that study of the natural world was way more important than study of philosophy or theology per se.[45] Partly for this reason, he and the Monistbund proposed radical changes to the educational system in Germany: toward more study of the natural world, and toward the study of historicized understandings of religion and theology. Whatever else religion and theology and philosophy might be, it must be attuned to the emerging evolutionary understanding of the world, and not vice versa. Religion, theology and philosophy ought not to straitjacket "nature," but nature should provide the contexts for human thought. This reattunement to the evolving natural world also meant that one should practice a certain multiperspectivalism about knowledge: If all knowledge is embodied, and embodiments are different (which one learns studying the natural world), then multiple perspectives about the world can give us a better understanding of our world than any one perspective alone. He saw any type of dogma as a hindrance to this type of multiperspectivalism and, as such, worked hard to root out that which would hinder epistemological progress. For Haeckel, "The goddess of truth dwells in the temple of nature, in the green woods, on the blue sea, and on the snowy summits of the hills—not in the gloom of the cloister, nor in the narrow prisons of our jail-like schools, nor in the clouds of incense of the Christian churches."[46]

It is true that Haeckel himself was certain that he was correct even though his theoretical assumptions contradicted such certainty. This, again, had to do with the progressive understanding of knowledge and time, and the direction toward which things progress that he imported from monotheistic theology. However, other theological influences encouraged him to turn toward experience, embodiment, and pluralism. He thought some theologians, such as Friedreich Schleiermacher, were on the right track.

Schleiermacher was a friend of Haeckel's parents, and his parents discussed Schleiermacher's thought with him often. He was intrigued by Schleiermacher's "turn to experience," and the idea that the fundamental experience of human life in the world is one of dependency. The main difference is that he would replace Schleiermacher's "God," with "Nature." For Haeckel then, the fundamental experience of being human is dependency upon nature.[47] As Richards argues: "Religious feeling, according to Schleiermacher's Über die Religion (*On religion*, 1799), might best be cultivated by seeking after truth, experiencing beauty, and contemplating nature. Haeckel practiced this kind of Schleiermachian religion all of his life."[48]

For Haeckel, whatever else religion or theology might be, it must fit the emerging evolutionary scientific narrative rather than forcing the material world into some immaterial (philosophical-theological) understanding. It was his opinion that one of the best "scientific" formulations of theology was made by Ludwig Feuerbach and his idea of religion as projection. Human understandings of God are collective projections, which has implications for how humans live with one another and with the rest of the natural world.[49] From an immanent framework, this understanding of religion as a collective cultural projection that has effects in the world is not necessarily a demotion, but rather just a naturalistic understanding of religion (not to mention ideas and other things immaterial). It was not, then, the case that Haeckel was a materialistic, reductive scientist arguing against religiosity, but rather that he was arguing for a new grounding for religion within the emerging picture of the natural world. He writes,

> Monistic investigation of nature as knowledge of the true, monistic ethic as training for the good, monistic aesthetic as pursuit of the beautiful—these are the three great departments of our monism: by the harmonious and consistent cultivation of these we effect at last the truly beatific union of religion and science, so painfully longed after by so many today. The True, the Beautiful, and the Good, these are the three august Divine Ones before which we bow the knee in adoration; in the unforced combination and mutual supplementing of these we gain the pure idea of God. To this 'triune' Divine Ideal shall the coming twentieth century build its altars.[50]

And, if Truth, Goodness, and Beauty served as a divine Trinity of sorts, then that Trinity can only be found in the natural world in all its ongoing manifestations. Haeckel worshipped at the altar of nature.

Haeckel's turn to nature and experience was not only theoretically based but also based upon the experience of traveling to other places. Before there was television or internet accelerating the process of global interconnection and knowledge, there were (of course) explorers or merchants whose travel had mainly political and economic power as goals; but they also included these earlier scientists (privileged to be sure) who had a sense of wonder about the diverse world and their goal for traveling was also to gain knowledge of the rest of the natural world (and not solely economic gain).

The fact that Haeckel went on several expeditions around the world in different bioregions helped him experience the diversity of human beings, human cultures, and the rest of nature in ways that were largely not possible much before his time. The turn toward immanence, embodiment, and experience of the wider planetary community allowed for the immense diversity within the natural world to creep into his thought and provided yet another reason to question what he thought of as the dogmatic slumber of theological doctrines. As the travels to Italy, Indonesia, and other places made clear, nature was much more diverse than humans had imagined; yet there were common forms and patterns that one could see within the natural world. He linked diverse forms into a continuum of life from abiotic crystal structures to plants, animals, and humans.

As already noted, his was also a progressive form of evolution, with (certain) human beings at the "top" as the most advanced, even if he saw the divisions between each "level" of life as fluid and not so clear as many would like to think.[51]

Thinking of humans as at the "top" of the evolutionary tree, which was synonymous with the most complex and most advanced or most progressed in the evolutionary process, was of course a way of thinking that smuggles in the Ptolemaic world of the Greeks, Romans, and later Christians and Muslims: with ultimate reality, forms, the unmoved mover "at the top" and materiality and embodied things at the "bottom." Obviously, Haeckel did not still believe in Aristotelean-Ptolemaic cosmology or biology, but this just goes to show the staying power of deep seated cultural ideas: even within an idea of an evolutionary model, the tree with humans "on top" as the reasoning creatures, or as the most progressed, holds to these Aristotelian metaphors for cosmology and biology (all teleological for Aristotle, of course).

As Haeckel began to see patterns in nature, and through an evolutionary lens began to break away from Aristotelian concepts of biology (though not wholly), he also sought organization of the differences among human languages and cultures. The various religious traditions, cultures and languages, seemingly incommensurable, were in need of ordering. It seems he followed the philologists and those promoting a "scientific

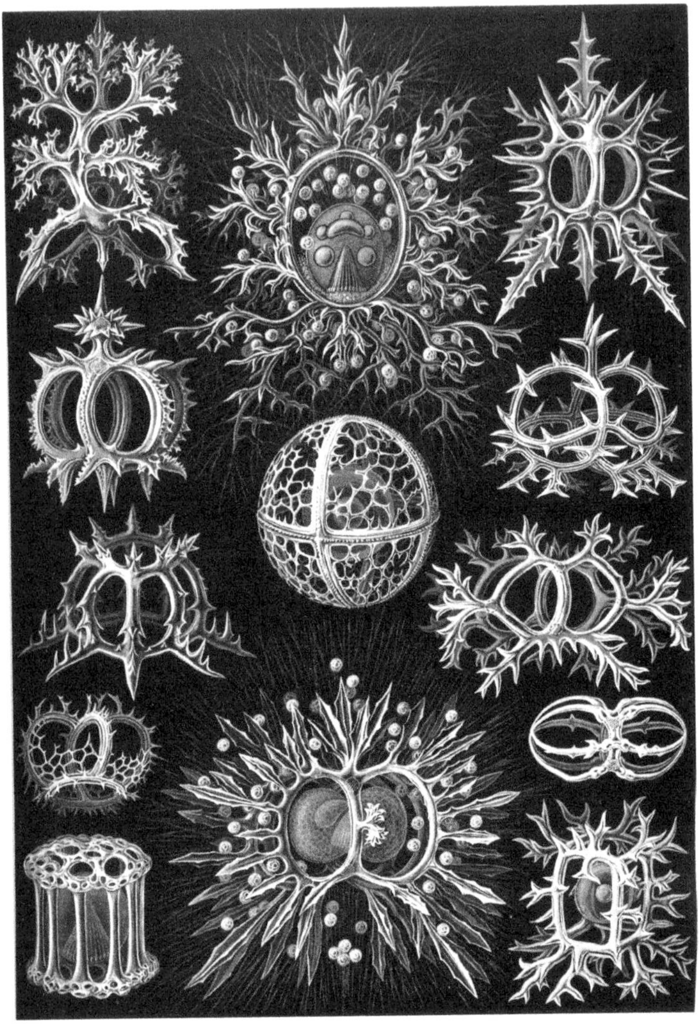

FIGURE 2.2 Ernst Haeckel, "Stephoidea" (plate 7), in *Kunstformen der Natur* (1904). Haeckel understood radiolaria to be the link between abiotic and biotic entities, which were all on a spectrum rather than disconnected. (Courtesy of Wikimedia Commons)

understanding of religion," that places languages, cultures, and religions into a progressive order leading up to the monotheistic cultures of Europe.[52] Whereas languages were seen as evolving toward more and more complexity based upon their grammar, with German being the most advanced, religions were put on a scale from animism to monotheism. The mono- of monotheism and (and monism for Haeckel) allowed for diversity and connection, but only through metaphors of unity and progression. As Kate Rigby and others note, "Romantic holism was hierarchical in orientation and almost universally privileged humanity, or more specifically, the human mind" and led to some form of "the divination of humanity."[53]

This reading of nature as single and progressive covered over its own roots in monotheism. From the perspective of one true nature (formerly one true God), the diversity of life is ranked toward the most advanced, humanity, and the multiplicity of human cultures were ranked according to the measure of Haeckel's own German, Western scientific, European location. This was a type of multiculturalism hierarchically ranked from the position of the center/top.[54] One big contradiction is that he places the heirs of monotheism "at the top," while promoting a new religion much more in line with animisms and pantheisms. These contradictory conclusions were theological materializations in Haeckel's own thought: further examples of how religion and science, humans and nature, material and ideas are always and already together and coevolving with one another.

His understanding of nature, like many others, had other indirect influences from theological understandings of reality. Concerns with origins and teleology that were taken up into early Christian theology quickly became translated into a single arrow of time, with a beginning (ex nihilo) and an end (telos/*eschaton*/world to come). This timescale was progressive and single. This progressive history was being strengthened by the early study of philology in Haeckel's time. Haeckel was in tune with the latest philological studies of the nineteenth Century which had begun to form an Aryan link between Indo-Germanic languages. This connection, in short, as Walter Mignolo notes, helped enable Christianity (and "white" Europe in general) to deal with the "problems" that Jews and Muslims posed to an otherwise linear timeframe that followed the torch of progress from the Greeks and Romans to Christians (and Latin Christians especially) and eventually to the Enlightenment and Western science.[55]

Making connections with Sanskrit, and especially Buddhist, thought, scholars were able to decipher connections and commonalities between India and the Ancient Greeks. In doing so, the "narrative of progress" was able to claim that Christians alone were true heirs to the Indic-influenced ancient Greeks (and to their capacity for reason) while Jews and Muslims (speakers of Semitic languages) were corruptors of this lineage. In other words, the historical connection between Indic cultures and ancient Greek cultures, and the idea that it was the Christians that took on this historical torch of (now) Western philosophical reasoning, made it easier to see Jews and Muslims as lesser than or corruptors of the one true narrative, reinforcing the so-called Dark Ages narrative.

In (unconsciously?) adopting this understanding of time, Haeckel swallowed whole a theological dogma that argued for one universal history, one story, and one truth; and a single history that was both antisemitic and Islamophobic. If this is one's model, then like earlier Christians before him, everything must be converted into this model's timescale, or ignored, or be gotten rid of. Thus, Haeckel's own form of naturalism, triune monism, had many elements familiar to monotheists.

TRIUNE MONISM

Haeckel was developing his naturalism during the Industrial Revolution. Industrialization, as Merchant and others have argued, required adopting a mechanistic worldview that understood most of the nonhuman world (and some of the human world—slaves and workers) in terms of what resources they might provide for the projects of (Western) human cultivation and civilization. Many in the field of environmental history and "religion and ecology/nature" have also argued that a mechanistic worldview and the understanding of humans as exceptions to the rest of the natural world were aided by certain interpretations of monotheism and the Genesis narratives. By the eighteenth and nineteenth centuries, many scientists, influenced by Romanticism and the problems that arose from industrialization, began questioning both the monotheism and the reductive model of the natural world.

For Haeckel, such a reductive model of the world could not be possible because it just recapitulates the dualism assumed by dogmatic theology. The theology that makes humans exceptions to or above the rest of the natural world, and suggests that goodness, truth, and beauty are transcendent, also makes the rest of the natural world "stuff" that can be used toward human ends. Though scientists increasingly understood humans as part of nature, most still held on to a special place for (some) humans as above and distinct from the rest of the natural world. In this way, a type of dualism seeped into mechanistic understandings of nature and materiality. Descartes's "ghost" in the machine was one example, as was Kant's *ding an sich* (again, at least according to Haeckel's interpretation of both). These types of dualism were just recapitulating years of wrong-headed theological thought, according to Haeckel. In place of such dualism,

Haeckel proposed his triune monism, consisting of energy, matter and "feeling/sensing." These three elements were present in life "all the way down" and this meant that the entire world could not be reduced and divided in the same sort of mechanistic, hierarchical way the Industrial Revolution demanded. There is sensing and value even in the abiotic, which bucks the idea that all life can be reduced to mere instruments toward certain human ends. From the abiotic to the biological, from the physical and chemical to the cosmological, energy-matter-and-"feeling" are present in various forms.

Whereas Newton had largely pushed out the need for God in understanding the heavens through his mechanization of the cosmos to no more than an originating force or point, and Darwin (et al.) had pushed away the need for God and soul with his evolutionary theory and the descent of "man," Haeckel wanted to ensure that we could do away with God-talk when we talk about origins and forms, meanings, values, and truth and beauty in the world. These were not properties of a soul endowed by God or any other ethereal realm, but emergent from the processes of evolution. Whatever was meant by soul (and consciousness), it could not survive the process of death and new life, because individuals emerged in the process of reproduction within species, in relationship to other life systems, and from common ancestors over time, not ex nihilo: individual organisms cannot be removed from their relationships. In this sense, forms morph, so what type of soul could possibly stay the same in light of such morphology? Haeckel argues,

> Our human body has been built up slowly and by degrees from a long series of vertebrate ancestors, and this is also true of our soul; as a function of our brain it has gradually been developed in reciprocal action and re-action with this its bodily organ. What

> we briefly designate as the "human soul," is only the sum of our feeling, willing, and thinking—the sum of those physiological functions whose elementary organs are constituted by the microscopic ganglion-cells of our brain.[56]

Note here that he is not reducing everything to mere material, but argued that the emergence of things like feeling, willing, and thinking allow for the emergence of something like the soul.

One of Haeckel's main critiques of the Christian project was its form of anthropocentrism. Again, developments in physics, chemistry, embryology, zoology, ecology, botany, and cosmology would help to correct that "anthropomorphic" projection of a God in human form by showing that humans were evolutionary products of the same processes as all other life. However, the answer for Haeckel and other Romantic scientists was not reductionism: if humans were a part of this process, then a reductive model would also reduce humans to mere resource. He writes,

> The monistic idea of God, which alone is compatible with our present knowledge of nature, recognises the divine spirit in all things. It can never recognise in God a "personal being," or, in other words, an individual of limited extension in space, or even of human form. God is everywhere. As Giordano Bruno has it: "There is one spirit in all things, and nobody is so small that it does not contain a part of the divine substance whereby it is animated." Every atom is thus animated, and so is the ether; we might, therefore, represent God as the infinite sum of all natural forces, the sum of all atomic forces and all ether-vibrations.[57]

This reading does suggest the universal "time" aforementioned, which ranks all of life in a hierarchy according to culturally located readings of nature, in a way divinizes some version of the human mind in the process. Despite this, he really did see humans as emergent in a common process of planetary evolution. His understanding of monism had qualities of emergent newness and was not what today would be called reductionist. Many of these emergent ideas come out of his understanding of evolutionary theory and readings of Darwin.

Haeckel's triune monism (energy, matter, and "feeling") was a full-blown religious naturalism: he wanted to explain everything based upon his own understanding of nature through the sciences (both natural and humanistic). He believed that the scientific study of nature provided a better route to understanding everything—including philosophy, culture, psychology, and religion—than anything the earth had previously sprung forth. His concern was with how the sciences could aesthetically and reasonably provide us with a better picture of the rest of the natural world than could any idea of religious creation. His understanding of triune monism led him to a huge appreciation of the immense diversity of the natural world, as seen in all of his nature drawings. He was mesmerized by the sheer diversity of life in the world and the connections that made everything fit together.

Haeckel was making a religion out of the sciences of evolution (change) and ecology (relationality). Through his understanding of these sciences, he tried to make sense of the world (*re-ligare/legere*), taking into account as much difference and diversity as is possible and reattuning to the emerging new world accordingly. Still today, such reattunement to the worlds in which we live is a process of recognizing our own embeddedness therein. Our thoughts are through and through ecological, always and already, and thus thinking itself is an environmental act. At the same time, these sense-making projects will always fall short, be blind to historical-cultural-embodied limitations, and will only be "for a time."[58] We must, then, find ways to critically examine our ecological meaning-making practices that weed out the more parochial and biased observations and cultivate the more planetary ones. It is a process and a journey ad infinitum with no arrival "home" or at some far-off end point.

CONCLUSION

In his many journeys, and thousands of intricate drawings/paintings of plants, sea life, and embryos, among other things, Haeckel saw patterns that led him to develop ideas such as his recapitulation theory. In addition, he argued for what we would call environmental factors as the main drivers of change in evolution. The point here is that he thought about the planet, including its meaning and value, as an emergent and continuous process. There was no fixed past in a point at which all life was created. but rather organisms live and evolve together toward many different possible ways of becoming, even though for Haeckel there is still clearly a progressive tendency in life and a "right" or "better" way to become.

This freedom and openness was, for Haeckel, part and parcel of the scientific project: part of the ecology of knowing itself.[59] Unfortunately, though understandably, he is remembered mostly for the polemical battles he waged against theisms of all kinds, and for his racist, ablest, and antisemitic leanings in terms of the evolutionary tree, rather than (also) as an ancestor to contemporary religious naturalisms. He has been mostly "canceled" academically except as a cautionary tale for the dark places evolutionary theory and Romanticism can go. I'm not at all suggesting these tragic and horrible things about him be forgotten, but they should be remembered along with the things that we might see as positive contributions to human thinking and as possible contributions to a different path toward future becoming.

We are all multiple in our subjectivity, and we all have cultural-biological-historical biases beyond which we can't possibly see. Polemics that paint Haeckel as merely an evil Nazi scientist have been detrimental to both the science and religion discourse, and to developing some type of nonreductive naturalism, and what I am calling a CPR. The pitting of Haeckel as an example of science vs. religion has helped forge a gap between science and religion for much of the twentieth century in the United States. Furthermore, such polemics helped religious people, especially conservative Christians in the United States, to understand value as something that has nothing to do with scientific understandings of nature, and even to regard environmentalism and other "paganisms" as somehow anti-Christian or evil. Finally, though there are deep connections between Romanticisms and nationalisms, Romanticism does not lead inherently to nationalisms, particularly, as I will argue, when it is critical and tied to a planetary scale. Only with the emergence of fields such as "religion and ecology" and "religion and science" in the 1960s have these oversimplifications and errors been called out in any systematic way. In the next four chapters I begin to look at what we might be able to do with a Haeckel-style Romanticism, when we make it pluralistic (rather than monistic), critical or multi-perspectival (rather than assuming the superiority of one location/perspective), and planetary (rather than place- or nation-based).

II

RETHINKING BEAUTY (PLURALISM)

3

ART FORMS OF NATURE

Aesthetics "from Below"

In Romantic attempts to bring together the sciences of geology, evolution, zoology, embryology, physics and cosmology under the umbrella of a new, naturalistic worldview, aesthetics played a key role. Haeckel argued, for instance, that the guiding values for naturalism would be the old Greek trinity of goodness, beauty, and truth; yet this trinity did not originate in some transcendent realm, but rather emerged in the process of nature naturing. From crystals to the most simple, single-celled life in nature to "societies" of complex cells that made up plants, humans, and other animals, Haeckel found beauty, forms, and patterns throughout. Typical of his musings on nature's beauty is the following perspective:

> Whether we marvel at the majesty of the lofty mountains or the magic world of the sea, whether with the telescope we explore the infinitely great wonders of the starry heaven, or with the microscope the yet more surprising wonders of a life infinitely small, everywhere does Divine Nature open up to us an inexhaustible fountain of aesthetic enjoyment.[1]

Many of the connections he made between various plant and animal organisms, and between the various sciences, were depicted in a number of his paintings and sketches: whether sketching his beloved medusa (jellyfish) or comparing embryos of various species. Among the most well known of these sketches can be found in his book *Kunstformen der Natur* from 1904. He sought in these drawings to sketch out the similarities in forms across many very different and very diverse species, and in doing so challenged the dominant theological aesthetics of the Christian West at the time. This chapter examines how Romantic scientists challenged three primary aesthetic categories: that between biotic and abiotic things in the world, that between plant and animal life, and that between humans and the rest of the animal world. In doing so, they helped to free identities from anchors in a transcendent realm that dictated boundaries between male/female, human/animal/plant, and biotic/abiotic life. They reattuned our aesthetic sensibilities to the diversity of life-forms in the world around us. It was from within these curious, blurred borders and crossings that understandings of humans in the Western modern mindset shifted toward humans being seen as part of the rest of the evolving planetary community.

CHALLENGING THE THEOLOGICAL AESTHETICS OF CREATION

As has been explored in the previous three chapters, the German Romantics of the eighteenth and nineteenth centuries deeply influenced Haeckel, as did the emerging scientific story of "nature" coming from the likes of people such as von Humboldt, and the emerging sciences of geology, zoology, botany, and eventually Charles Darwin's evolution. The influx of knowledge about other cultures made possible via faster modes of transportation and the influx of knowledge about the natural world made possible by advances in the microscope (the small) as well as the telescope (the large) enabled closer examinations of the natural world and the differences found within and between humans and other life-forms therein. As Marsha Morton points out, "Nature [for Haeckel] was seen as a wonderland of new aesthetic enjoyment, provided by the telescope and microscope, in which poetry was immanent."[2] As for many others of his era, the story provided by theology and the church in Europe broke apart for Haeckel under the pressure of the growing understanding of geological time and the immense diversity of life-forms extant and extinct on the planet. The deep past was literally being uncovered in the fossilized rock formations of the present, and the knowledge being produced by the sciences was challenging the future presented by the church and opening up new possibilities for becoming on the horizon. The human location within the cosmos was in need of aesthetic reattunement.

The transcendent source of beauty, goodness and truth, was being questioned and with it the aesthetics of transcendence that ordered the world "from above." The hierarchy of God-humans-nature, and the dualism of spirit/matter (and along with it good/bad) was being undone. Haeckel and other Romantic scientists enter this world as this turning, this reattunement, is in full swing. As mentioned earlier, the turn to experience in Schleiermacher and his idea of religion, as dependency upon God turned (for Haeckel) into the dependency upon Nature, was important.[3] He was also very much influenced by the demythologizing of Jesus and the first historical Jesus scholars of the nineteenth century.[4] Finally the idea of "religion as human projection" found in Feuerbach (among others) seemed to make much more sense for Haeckel within the emerging naturalistic worldview he was constructing.[5] In all of these shifts there was a turning away from the transcendent and supernatural as the source for value, meaning and truth, and toward an immanent framework based upon what we could know through reasonable deliberation of data gleaned through observation of the natural world through our senses. Haeckel's struggle was that of articulating a shift in meaning, value, knowledge, and aesthetics away from the dualistic and transcendent theological model and toward that of an evolving, immanent, naturalistic model based upon the emerging scientific story of the planet and its life-forms (including humans).

What we might identify in his art works and in his writings as an emergent aesthetics of nature, was to him an attempt at explaining this monistic, evolutionary "scientific" view of the world, which navigated a third way between materialism and idealism. His aesthetic was not something found at extreme ends of ideological spectrums but in the everyday examination of life. In doing so, I argue, he ends up returning beauty and value to the world by replacing the transcendent monotheistic God with immanent Nature. It is an aesthetic reattunement to the world from within an evolutionary perspective. His version of monism returns wonder to the world through making the God of theism (or better deism) into the living Nature of monism.

> Insofar as wonder can function as a kind of wound in the everyday . . . it must again be emphasized: just as a wound ceases to be itself when it heals, wonder is only wonder when it remains open. Wonder opens an originary rift in thought, an unsuturable gash that both constitutes and deconstitutes thinking as such.[6]

Wonder (*thaumazien* for Aristotle and the Greeks) is that which, as Mary-Jane Rubenstein suggests in the above quote, holds open the space for life to spring forth or emerge. It is the refusal to allow our thinking to be a form of what Heidegger calls "enframement" and is instead an opting for poiesis.[7] The extremes of idealism and materialism lead toward an ideological distortion of the world and our knowledge of it, just as much as the imbalance of humors according to Hippocrates and Aristotle lead to illness in the human body. "Balance" or "a middle way" seems like a strange place to locate wonder and awe, but I want to argue that far from merely being a method for an enlightened, "reasonable and balanced" approach to understanding the world, the via media is grounded in wonder at the embodied natural world and this method is taken up by the German Romantics in a semicritical way, including Ernst Haeckel and his search for a naturalistic worldview.

Haeckel's "middle way" is not unrelated to the technologies of scale—the microscope and the very small/close, and the telescope and the very large/far away—that suggest a sense of wonder for the unfolding of life experienced by humans on earth as somewhat "in between." In other words, the source for value, wonder, and beauty is not to be based in some sort of *telos* written in to the rest of the natural world nor in some sort of transcendent form that orders the natural world; rather, value, wonder, and beauty are emergent aspects of the world of nature that should be discoverable through closer observation of nature and the life therein. It was in the "in-between" space that Haeckel's art and aesthetics became central to his understanding of knowledge. Knowledge cannot be enframed in a transcendent or foundational aesthetic, but rather it emerges and bubbles up. As Breidbach points out, the *Artforms of Nature* and other artistic works of Haeckel were meant to be instructional books for the common reader in how to see the natural world. People had to be trained to read and interpret nature from a naturalistic rather than a theological worldview. Evolution is the process by which we come to know the natural world, and we are a part of that very process. For Haeckel,

> humans are nature, they are a part of, and the result of evolution. Our actions and thoughts are products of this evolution. Accordingly, when humans come to know something, ultimately it reveals their own nature. Our knowledge—which has developed in and is subject to the laws of nature—is in itself nature (and according to Haeckel, nothing more).[8]

Knowledge is gained through experience, sensing, and interacting with the world around us, not by fitting data into a priori categories. Perceiving is knowing, but not in the sense that we have access to "bare facts," but rather because we are part of the process that is unfolding and thus have evolved in a way to perceive the patterns of the rest of the natural world that also make up humanity. Such perception is not a grasping or enframing, but rather an attuning to the life unfolding around us and constitutive of us. It is precisely this reason that objective certainty takes us away from knowledge of the world and holding up spaces of uncertainty, spaces of encountering others, brings us down to earthly realities. As Louise Economides notes, "Uncertainty *connects* the poet to such things because it reflects an ability to engage with nonhuman otherness in all of its strangeness, a refusal to reduce such phenomena to the comforting familiarity of place with all its potentially sentimental trappings."[9] And further, "Accepting the wonder of a groundless existence entails loving the other *as other*, not seeking to use the other as a foundation for transcendental 'truth' or reducing the other to a reliable means whereby subjectivity defines itself in opposition to what is not."[10]

From crystalline structures, to radiolaria, to Haeckel's beloved medusae, to plants, other animals and humans, similar forms and patterns can be found that link all the evolving life on the planet together. These forms and patterns, and the ways in which they get combined are, however, open-ended. There is no teleological understanding of nature for Haeckel, though he does believe that all organisms, even human ones, tend toward perfection within their environment.[11] Wonder at and study of the rest of the natural world would—through direct examination and through nature sketches and paintings—according to Haeckel, help lead to the perfection of societies; this was not just a rational study, but one that retrained the affects as well! As one scholar comments,

> Haeckel himself exhibited a messianic zeal in promoting his logical, new pantheistic, nature religion through lectures during which he would display his own beautiful hand-colored drawings and etchings of cells, embryos, and other natural phenomena that appealed on an emotional level to those seeking a greater meaning in life through the study of its apparent rationality, organization, beauty, and essential unity. It was visual material that had a striking 'shock of the new' quality about it in an age without cinema or television.[12]

Again, as Rubenstein notes in *Strange Wonder*, following Aristotle, at the base of all of our knowledge is not certainty but wonder. Wonder is that which keeps our concepts open to an ever-changing world, disabling a secure connection between thoughts/ideas and things, or language/concepts and entities in the world; it enables the world to be a living, mattering entity. It is this deep attention to the mattering of the world of everyday experience that brings Aristotle to a different epistemological method than that of his teacher Plato. Aristotle was working out a via media between the idealism of Plato and the materialism of Democritus and formulated a sort of hylomorphism in which the essence and material were related to one another. This hylomorphic idea, adopted and adapted by many over the last two millennia, would later be picked up by Haeckel in describing the relationship between matter-energy or material-ideas, ultimately culminating in his theory of recapitulation: ontogeny recapitulates phylogeny.[13] Going too far in either direction—erasing ideals from matter or erasing matter from ideals—trapped the world into either a function of instrumental scientific calculation, or a function of philosophical (or theological) dogma. Either way, both options are only from a human perspective and turn the world into that which can be understood (and later used) by humans.

This reading of Aristotle is, of course, one reading, but it is consistent with the ways in which wonder has often been associated not with the aberrant or exceptional, but with the habits, repetitions, and regularities of the natural world.[14] Many natural philosophers of ancient Greece, like German Romantics of the nineteenth century, sought wonder in what they perceived to be the everyday realities of nature naturing. In a way, the fact that there was order at all, that life kept on going, seemed to be the cause of more wonder than irregularities and so-called miracles that have been at times seen as "wondrous," if not fetishized into some sort of spectacle. We might say this is a reading from a third space, that necessary space of relationality, relating one thing to another and through which all identities are made possible.[15] Or, similarly, we might call this space of relationality a queer space: one implication of the theory of evolution, after all, is that in reality there are no real species distinctions; rather, species are shifting constantly throughout time and all life has common origins and all boundaries are fluid. In addition, placing humans and all other life into an evolutionary perspective means also undoing the ideas of fixed genders and sexes: just as there are a multitude of sex and sexual expressions in the rest of the natural world, so there are a multitude of sex, sexual and gender expressions within the human realm. Reality is the evolving flux of life; species, sexes, genders and the like are nomenclature and categorical tools.[16] Whatever we might want to call it, this "third space," "queer space," or "middle space" might also be thought of as the space of lived, present experience between knowledge of the very small and close (made realizable through biology, geology, and the microscope) and the very large and far away (made realizable through cosmology and the telescope).

Finally, the temporality of this "third space," where wonder seems to bubble up in the everyday, might also be understood as a turn away from tradition and the past (theology and culture) and a freeing of possibilities for the future (through scientific knowledge), both of which allowed for a focus on the present and our experiences of the present. Rather than our minds being captivated merely by transcendent "far-off" or "distant" things, ideas, and places, the goal was much more to focus on the here and now.[17] Whatever else the "third," "queer," or "middle" space signifies, it is meant to signify the place of the living grounds of relationality that refuses reification and totalizing epistemological claims. Staying with this space, "staying with the trouble," enables us to pay deep attention to the shifting pluralities that make up our embodied realities.[18] It helps us to stay within what Bruno Latour and others referred to as "the critical zone."

> Learning to pace up and down the critical zone doesn't mean either going backwards, or forwards to the here below of days gone by, or towards the material world that the Moderns wanted to make the most of while spurning it to escape elsewhere. We can't escape anymore, but we can inhabit the same place in a different way, and this means all acrobatics, as Anna Tsing would say, have now to be based on the new ways of *placing ourselves differently* in the same spot.[19]

Haeckel worked within this present, immanent, middle space to articulate his understanding of monism and nature aesthetics. He marveled at the enormous diversity of life-forms, and the similarities and patterns he saw throughout nature, even recognizing a diversity in sex and sexuality as his influence on Magnus Hirschfeld's work shows.[20] Haeckel realized that evolution meant the uncoupling of identity from foundations that were laid out in some teleological or blueprint order by God. He knew this would mean an increase in the diversity of what counted as possible ways of being in the world. Evolution meant the end of God-created order, and this for Haeckel was good news and freeing. Freeing human thought from ideal forms or foundations enables humans to reattune to the worlds bubbling up around us, rather than conforming to those preconceived worlds. It was only through observing nature from this "third space" that the poiesis (or self-organizing emergence) of life could be really understood and respected.

However, there is no zero-point thinking. This would be an acontextual form of thought and thinking. In Haeckel's case, thought became tied to the imaginary context of the German nation and had disastrous consequences. I'll say more about this in the next chapter. My argument here is that if all thought is contextual, we ought to think about the scale of that context.

Further, if we tie a critical romanticism to the planetary imagination (the scale of the planet), we might prevent these parochialisms and nationalisms as well as human exceptionalism.

Haeckel's monism sought to navigate a way out of reductionistic and dualistic thinking, which cuts us off from the source of beauty, goodness, and truth that could be experienced in the natural world. The Industrial Revolution had, by Haeckel's time, begun to co-opt many of the natural sciences into a productionist model, and it was Haeckel's view, along with other Romantics, that the true goal of the natural sciences was more about instilling a sense of wonder about the world, and our place therein, than it was about technology transfer. In other words, science was more about aesthetics and value than production. As Olaf Breidbach notes, for Haeckel, "Knowledge of nature is 'natural aesthetics.' Accordingly, aesthetics are nothing more than reflections of nature itself. Nature, which develops out of and into itself, is beautiful."[21] Matter was not, for Haeckel, just dead stuff that was somehow fodder for human ends, but it was all a living, evolving entanglement that included human beings and human cultures. Whatever "spirit" or "soul" might be, it had to exist "all the way down" even if in simplistic form. For Haeckel, "the symmetry of forms that fit together into a unified nature aesthetic, is evidence of the ensoulment of everything that exists."[22] Little did Haeckel know he was following in line with early church fathers and Neoplatonists that had some notion of a world soul, or the ensoulment of all life.[23]

Haeckel's monism is not one that does away with pluralism; as stated before, Haeckel's substance had a triune structure to it (matter, energy, and subjective experience). There was room for diversity in Haeckel's monism, despite some of his more shortsighted nationalistic ideas and his belief that science would eventually solve all mysteries of the world. His method still depended upon that middle space of wonder, which refused closure. It allowed for the wonder of the natural world to come into view with eyes anew: in short, undoing, and unknowing enabled a form of poiesis that allowed life to appear in new ways toward which we might reattune.

More than autopoiesis, I would argue that he articulated a form of sympoiesis of the kind that Donna Haraway writes about in *Staying with The Trouble*, and which F Schlegel coined way before that.[24] Haraway argues that we should use

> sympoiesis for [describing] "collectively-producing systems that do not have self-defined spatial or temporal boundaries. Information and control are distributed among components. The systems are evolutionary and have the potential for surprising change." By contrast, autopoietic systems are "self-producing" autonomous units "with self defined spatial or temporal boundaries that tend to be centrally controlled, homeostatic, and predictable.[25]

It is not just that life was self-created, but that life-in-connection-with-life worked together to create the community of life. Haeckel's theory of recapitulation disables, in a way, the idea that a single organism can self-create: it is created by all that has come before it plus the surrounding *oikos* into which its own unique life is thrown. Nature can provide us our forms, our values, and the grounds of our being, but not in any naively romantic way. Because each life is "checked" and "chiseled" by the rest of the communion of life it finds itself within, there is no direct knowledge of the world: we only know the world through communion with it and communication about that communion, each of which depends upon the multiple perspectives in the evolving planetary community that make up a given moment. This is a far cry from the individualistic interpretation of the Genesis account of humans as male and female created in a special place above the rest of the natural world. It is also a far cry from the individual struggling and competing for resources read into the evolutionary story of the world. This type of critical romanticism, then, is the outcome of placing ourselves within the rest of the planetary community, through an evolutionary story that is not tied to any guiding ideology other than that we inhabit a common planet and life "wants" to continue. We cannot have thinking without context, but if we hitch our thinking to human exceptionalism or nationalism, it distorts the planetary community toward that context. Again, the planetary is hopefully a large and open enough imaginary for thinking about how we might live into a planetary future more focused on the flourishing of all its members.

I'm not arguing that Haeckel was an apophatic, postmodern thinker. He was, after all, a scientist looking for better understandings of the way the world actually is. What I am arguing is that Haeckel the philosopher and artist, among other Romantic scientists of his era, recognized that received knowledge was wrong, and that a new method for understanding the world should have built in mechanisms for allowing knowledge to be challenged and changed.

What Haeckel was clearly arguing against was transcendent metaphysics that for him was the language of certainty that closes entities off from the world, creating certain objects in the world, and the world eventually into a certain object. This closure of metaphysics is of the same type that Carolyn Merchant describes in her now classic *The Death of Nature*. The closure of chance and wonder in the natural world through mechanization was, in part, the result of bringing Newtonian mechanics to bear on all of the natural world (geology, evolution, physiology, etc.). Even though for Newton this was not the goal, his mechanics was transformed from metaphor to actuality, which were manifested by the bureaucratic and mechanistic nature of the Industrial Revolution. These mechanizations were also manifested in political and religious dogmas of Haeckel's era.

From this perspective, it is no surprise that Haeckel (like Schelling before him) draws from Bruno in developing his own version of pantheism (who himself drew from apophatic theology), and even places his own intellectual work in continuity with Bruno, Copernicus, and Luther, in arguing for a reformation in the ways that we understand the world.[26] What all of these thinkers practiced, to varying degrees, is what might be called the art of unknowing. In contemporary queer theory, Jack Halberstam has linked what I would call apophatic moments of realization with "the queer art of failure."[27] These thinkers were engaged in purposefully failing their performances of the received traditions of wisdom and failing to take orthodoxy as an answer to life's riddles. This doesn't mean they saw the world from an objective standpoint, but it does mean that these failures, these spaces of unknowing enabled a sense of wonder to reignite curiosity about the world with new possibilities for becoming. This failure enabled opportunities for reattuning to the world in different ways.

Haeckel's own version of unknowing can be seen in his refusal of human access to bare facts in the world, a refusal which many scientists of his time (not to mention our own) did (and do) not like.[28] This "uncertainty" is simply the result of insisting that history is a part of evolution and geology.[29] From this perspective, one could argue that placing humans within an evolutionary framework also implies that whatever we mean by *beauty*, *value*, and *truth*, they must be emergent understandings of these terms. However, such an emergent understanding can also slip into an infinite regress that becomes relativistic. In a way, Haeckel's own aesthetic effort to prevent such a "slip" can be found in his understanding of ecology.

ECOLOGY AND POSTMODERNISM

Ecology, as Haeckel saw it, was the science that brought together the particular organism in relationship to a community of organisms. His theory of recapitulation (ontogeny recapitulates phylogeny) itself is also a product of what I have called his via media approach to the natural world. What he argued was that every individual goes through the history of "predecessor" organisms in its own development. The environment and context of that organism is what forces evolutionary changes. According to Haeckel, this environment and context for humans also includes language and culture. Ecology, then, becomes the context and aesthetic container for understanding the world around us thereby preventing a slip into infinite regress: If we can get a better understanding of the overall ecosystem into which we are born, then we can bring about the best, most beautiful, and truest life possible. This ecological context does provide firm grounds, but not eternal foundations: Ecosystems do change over time and organisms do evolve. This is why the Romantic position must always be critically romantic. Furthermore, if ecosystems drive evolutionary changes, and there are multiple ecosystems, then it suggests a drive toward pluralism, diversity, and even, in a sense, many worlds that together all make up the planetary community at any given time. In other words, there cannot be just one interpretation of "nature," and one standard by which we can judge all of nature: The planet is through and through multiple. Yet, it is also in

Haeckel's understanding of ecology that he became too closed off from the evolving source of life and thus where he fell into some problematic thinking. The closure within Haeckel's thought becomes clear in his ranking of differences rather than an acceptance of the diversity of humans as just different ways of being/becoming in the world.

Despite these very serious problems, I argue that Haeckel and other Romantic scientists can offer an alternative basis for a naturalistic, scientific worldview to that of the reductive, productive, instrumental model. It is also one that acknowledges common grounds, without claiming universality or foundations. The many ecosystems that make up the planet at any given moment are also entangled and connected. Humans also experience worlds differently depending upon our upbringing and embodiment, yet we also have common planetary grounds. One might think of gravity, or the need for earth, air, fire, and water. The point is that there are many different ways to co-construct worlds within the planet that don't reduce the planet to the perspective and needs of some people, such as the productive, reductive model does.

One of the keys to this alternative, nonreductive naturalism is found in the boundaries he challenges with his nature aesthetics (mentioned above). Now that I have fleshed out some of the context for his "aesthetics from below," I spend the rest of this chapter examining some of the ways in which his art challenged (and still challenges) some long-held boundaries within the natural world. Through challenging the boundaries between inorganic and organic, plants and animals, and humans and the rest of the natural world, I argue that Haeckel's nature aesthetics provide a resource for contemporary religious naturalisms and contemporary aesthetics of immanence in general, one that I am calling a critical planetary romanticism (CPR) for the earth.

Challenging the Boundary Between the Living and the Dead

Central to understanding Haeckel's nature aesthetic is his understanding of the origins of life. For Haeckel (as for Goethe and Erasmus Darwin among others before him) all life emerges from the sea. This was an important move in the eighteenth and nineteenth centuries: from the depiction of an idyllic garden as the origin of life, to that of the mysterious and even chaotic sea.[30] Such a change marked also the move from a created order to an idea of life evolving (somewhat chaotically and haphazardly). This reading excited Haeckel (and others) because it had the potential to move an aesthetics of creation from the hands of a creator god (above) to that of an emergent process in the natural world (from below).

The process of moving from the abiotic to the biotic follows the story, not of a creator god, but of the spontaneous generation of monera, for Haeckel. Again, this spontaneous generation or sympoiesis was the creative transition from life to nonlife. These monera "arose from the primeval ocean through the operation of purely physical and chemical conditions and were the origin of all living forms."[31] Monera were single-celled organisms, made up by an energy event which brought together "albuminous combinations of carbon."[32] They contained merely one particle of the protoplasm, which Haeckel believed was the source of what we might call an internally driven "life." Protoplasm is the soul/energy that is found in all living things. Though even in abiotic life, there is a precedence for biotic life: as Haeckel's *Kristallseelen* (Crystal Souls) book toward the end of his life makes clear. It is merely the distinction between self-organizing life and life which is acted upon purely by external factors.[33]

Prior to the formation of monera, the development of life was more akin to crystalline structures, which is one reason Haeckel pays so much attention to the patterns that make up crystalline and abiotic structures: for him these patterns are also found in biotic life, except each individual living organism begins to "remember" these patterns and forms and through them attune to the worlds in which they live—via the protoplasm—and pass those "memories" on to their progeny. The transition from abiotic life to biotic (ensouled) life, then, is more a transition of the locus of organization of a life than anything else. In other words, whereas rocks and crystals and other forms of abiotic life are "shaped" by the environment and climate around them, beginning with monera, entities also begin to react to and interact with their surroundings.[34]

With monera, the *umwelt* is internalized and becomes the inner world of the living organism. The living organism then becomes part of the *umwelt* in which it exists and is both an actor and acted upon. One is reminded here of the Daoist micro-macro/cosm in which the micro is a manifestation of the macro.[35] However the micro cannot be subsumed by the macro, and the macro cannot be reduced to the micro. In other words, individual organisms and the other biotic/abiotic world around them cannot be totally separated from one another; rather, they are continuous. This also means, among other things, that life is continuous with nonlife, and that physics, chemistry geology, and biology can all begin to fit together into a planetary story: a story in which life, energy, soul, sensing, and eventually thought and consciousness can be explained within an immanent framework.

To my mind, nothing captures the movement from abiotic to the biotic better than Haeckel's drawings of radiolarians. These are monera in which the protoplasm has come together

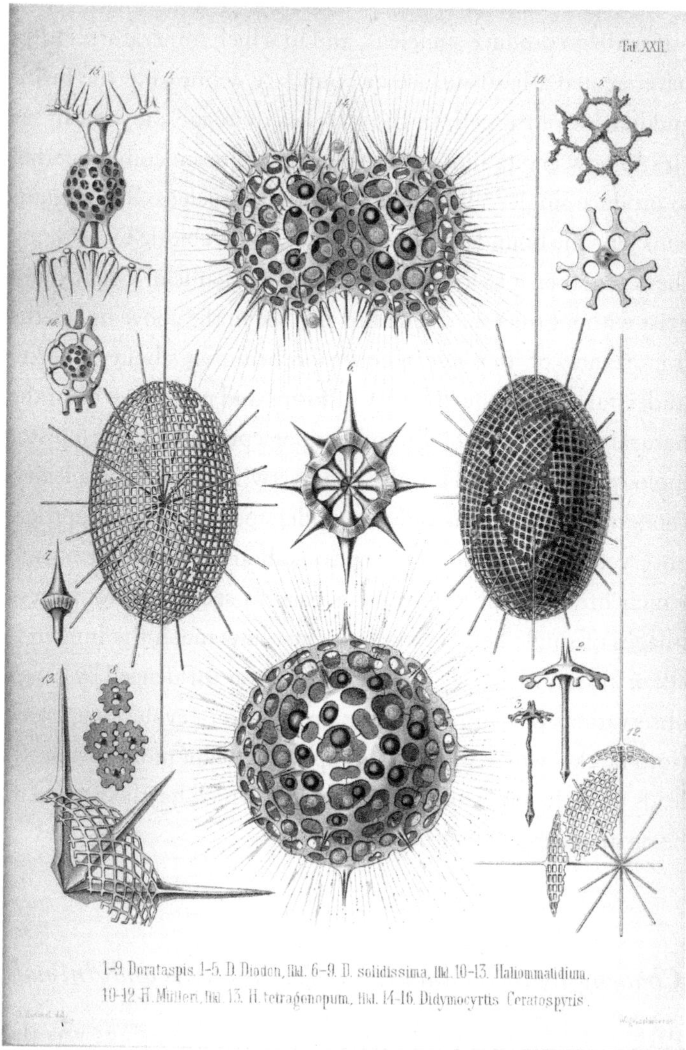

FIGURE 3.1 Ernst Haeckel, "Tafel XXII," in *Die Radiolarien* (Berlin, 1862), www.biolib.de/haeckel/radiolarien/index.html.

internally to produce a nucleus, and in which mineral structures have created a hard surface or boundary delineating the inside and outside of the cell. In the sketches in image 3.1, you can see clearly how crystalline patterns and protoplasm come together to produce single-celled organisms with a nucleus. Such organisms blur the boundaries between the biotic/abiotic by bringing them together into an emergent, single organism. One cannot exist without the other. One can also see in the above image the importance of the microscope for bringing about a better understanding of the diversity of forms and patterns within the natural world. Technology and our perception of nature (epistemology and ontology) are always intertwined. Such complexity, thought Haeckel, can only be understood through empirical observation: not a priori assumptions about "created" forms into which life must fit. Again, it is easy to see, even from these images of single-celled radiolaria, just how much the Jungendstil or art nouveau style influenced and was influenced by closer observations of nature. Moving up Haeckel's evolutionary tree (for Haeckel where was still an "up" that meant more advanced) leads us to the plant and animal worlds, which he understood to be on an emergent continuum.

Challenging the Boundaries Between Plants and Animals

Part of what Haeckel wanted to show in his drawings was the continuity between forms in nature, and the ways in which these forms combine to form new organisms. In the language of emergence, forms (a crystalline structure and single cell) come together to produce something else from nothing but.[36] Depending on how one divides the forms, or structures, one sees different aspects of nature or a given organism. One can, for instance, map the cells, tissues, and now genetic and chemical layers of the organism but the organism only comes into view with all these layers working together. One reason that Haeckel preferred drawings and sketches of nature to the emerging technology of photography was because of this ability to distinguish various layers and textures. For Haeckel, photographs took something away from the multiperspectival way in which a single organism might be observed: from different angles, at different depths, inside and outside, etc. This embeddedness within the natural world was strengthened by his sense that all biotic life shared in some level of "soul" (through the protoplasm), and that all forms also evolved.[37] There was, for Haeckel, no fixity, telos, or container by which the natural world could be made to appear to us as it really was, even if he did adhere to some sort of "progressive" elements overall in evolution.[38]

As Mike Hogue argues well in his book the *Tangled Bank*, Darwin's *Origin of Species* and *Decent of Man* were about the recognition that there are no unchanging forms in nature: rather forms also evolve.[39] In other words, there are no distinctly created species. As one of Haeckel's most cited quotes argues, everything is transformed sunlight. This means that the hermeneutic task is central to scientific investigation and paying attention to the many "voices" of nature is important. If everything that existed, exists, or will exist is in some way immanent, everything is interrelated, and a plurality of forms exist, then taking into account multiple perspectives of the worlds in which we live gives us a better understanding of the overall planetary community of which we are a part. Again, following other Romantics such as Goethe and Schelling, for Haeckel, there is no "outside" view, only multiple views from the "inside."[40] This thinking also comes very close to the insight of Deleuze and

Guattari in the twentieth century that the virtual is the real.[41] The ever-changing flux of embodied life is the really real, our fixing of forms is (as Whitehead notes or Buddhists might argue) misplaced concreteness or an illusion.[42] This opens up our thinking toward multiple forms of agency and actants that we find in many forms of the so-called new materialisms, or which was long before found in various forms of animisms, pantheisms, and panentheisms.[43] It also means that nature is through and through queer: boundaries and identities shift, and distinctions blur into one another.

Because of this flux in forms, Haeckel was interested in what we might think of as transitional forms of nature: those that blur the boundary between living and dead (discussed above) and human and animal (discussed below). Perhaps no animal was more studied for this transitional reason than his beloved medusae (jellyfish). Jellyfish, for Haeckel, represented the shift from plant life to that of animal life. Its reproductive cycles, though varied, often involve polyp-style offspring that seem more plant like (and attach to other sea life), which then grow into free floating organisms. Who is to say where a single jellyfish ends and begins: the life cycle is continuous; matter is continuously transformed into these elegant, magnificent creatures; in a sense these creatures are among the only immortal ones, challenging then also the boundaries of life and death, and the theological hierarchy of creation that place immortality above change and decay as an attribute of the gods, and not of "lowly" creatures. Their behaviors also blur the boundaries between plant and animal life: acting in similar ways to both types of life. Trickster figures, if there ever were such a thing!

In figure 3.2, one can see different views of the internal structures that make up this particular jellyfish on the periphery of the sketch. This is typical of Haeckel's embedded

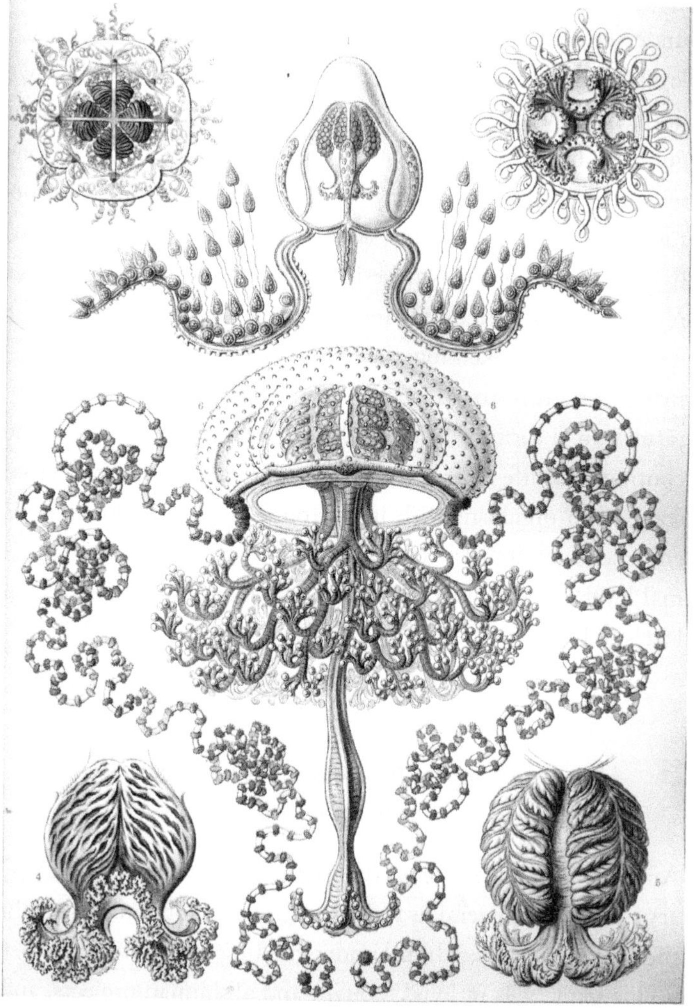

FIGURE 3.2 Ernst Haeckel, "Anthomedusae," in *Kunstformen der Natur* (1899). (Courtesy of Wikimedia Commons)

understanding of observation: how one observes and at what level of life one observes makes a difference in terms of how we understand what an organism is and the world around us in general. The two central structures in the sketch provide us with the more plantlike features of the jellyfish, while the bottom two structures begin to sketch out part of the more animal (tissue) like structures of the jellyfish. The top two structures reveal Haeckel's desire to find symmetry and order within the natural world. Taken together, this plate represents the micro-macro structure that Haeckel captures in his theory of recapitulation (a form of hylomorphism). It also reminds us of the continuity of life and the continuous interaction of the internal state of an organism with its surroundings (or ecology). This embeddedness and relatedness did not stop at the plant and animal levels of life for Haeckel, but extended also to the human levels of life.

Challenging the Boundaries Between Humans and Animals

Another aesthetic move on the part of Haeckel, and evolutionary thinkers in general, was the reattunement of humans within the evolutionary tree. In his understanding of evolution, history is emergent, as is rational thought and all human projects, and there are predecessors in other forms of life. For Haeckel one of the worst parts of monotheistic understandings of the world was, following Feuerbach, that humans had merely projected the "Human" writ large as the center, creator, and sustainer of the universe. This "gaseous vertebrate animal," as he called it in his more combative debates with theists, was the greatest delusion that kept humans from understanding their embeddedness with other animals in an evolutionary community.[44] In other words, Haeckel argued in a way that "it takes a planet to make a human," or at least a long process of evolving organisms and ecosystems.[45] He was opposed to transcendent hierarchies and dogmas that built up an idea of humans as exceptional and he saw it as crushing what we might call the planetary awe at the base of our deep relationality. Projects of human meaning, rationality, and consciousness writ large become a container into which all other life must fit. In other words, the monotheistic and transcendent basis for humanity played a role in reifying the world around us and thus in distorting the beauty and wonder that emerge from the rest of the natural world. Though Haeckel was a product of his own socio-historical location and would read that space back into the natural world as "Nature," he at least began to question the idea that human thought can exhaust the knowable possibilities of the world.

Again, Haeckel was adamant about (re)placing humans and everything human within the natural world. In figure 3.3, for instance, you can see both the embryonic development of a human, a ram, a bat, and a cat, but also the sketches of the final form of the faces of these animals. As, Marsha Morton points out, the choice by Haeckel to use a Greek god's face, was a deliberate attempt to transplant the origin of humans—and all things human—from anything theological natural to something like natural philosophy.[46] He sides with natural philosophy here as found in the Greeks, in other words, rather than anything biblical. Again, he seeks to strike a balance between various life-forms (the individual animal faces) and the patterns that hold them all together (found also in the face, but more so in the embryonic similarities of the various species). Such an immanent frame for understanding history, culture, art, politics, and all things human meant that something like Kant's

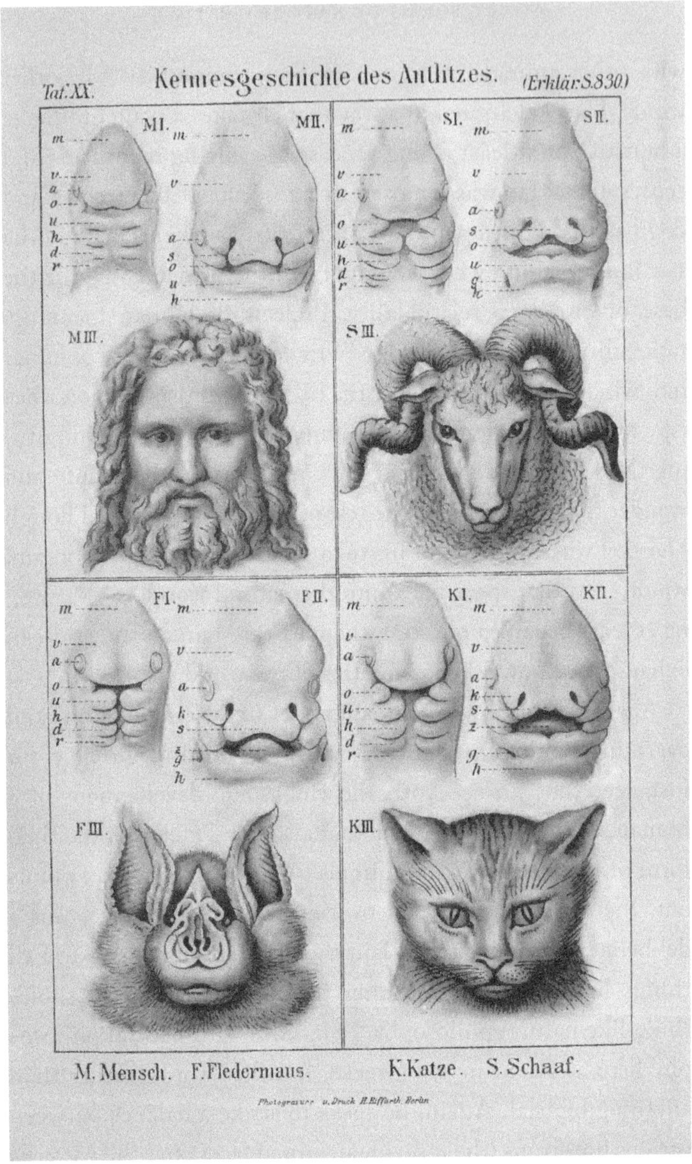

FIGURE 3.3 Ernst Haeckel, "Tafel I: Entwickelungsgeschichte des Gesichts," in *The Evolution of Man: A Popular Exposition of the Principal Points of Human Ontogeny and Phylogeny. From the German of Ernst Haeckel*, vol. 1 (D. Appleton, 1897), www.biodiversitylibrary.org/item/24127.

transcendental philosophy, Hegelian idealism, or the dogmatic theology of the Catholic Church—which all suggest that there is more than the emerging, changing natural world—were, in fact, distortions of truth, goodness, and beauty.

An immanent beauty and wonder, he thought, would give us a meaning-filled understanding of our place in the world that would transform our worlds for the better. In this sense, we can say, that Haeckel's (mostly) nonteleological (or at least emergent teleological) and nonfoundational understanding of nature meant that there was genuine freedom for becoming in different ways together as a planetary community.[47] The evolving community of life, for Haeckel, was pluriform. Every organism or entity reattunes to the worlds of which it is a part. The surrounding organisms and entities that make up the environment (ecology) of a given world lure organisms and entities into tune. As these ecologies are always in flux and changing, life is a constant reattuning process. This is not some sort of monological unity in which everything becomes some kind of a harmonious one, but it is a plurality, or a multiplicity constantly reattuning to one another.[48] In humans, this is a constant remaking of meaning and reimagining that is at the etymological heart of *religion*. For this reimagining, we need new aesthetics and meaning-making practices that are wonder-filled and that keep us focused on the everyday embodied becoming plu-realities of the planetary community and how these ideas, practices, aesthetics and values effect the various embodiments of transformed sunlight that make up our lives. This recovery of Haeckel and other Romantic empiricists of his era for a CPR is the topic of chapters 6 and 7.

CONCLUSION: A CAUTIOUS POSTMODERN AESTHETICS OF NATURE

As the reader may be aware, Haeckel's monism, the shift toward an evolutionary framework, and German Romanticism were put toward negative ends with the rise of National Socialism and the Third Reich in Germany, and in eugenics programs in the United States and Europe in general. The negative possibilities, the ugly side, of such a "nature-based" aesthetic, philosophy and ethics should never be forgotten. However, I argue that this negative side is also due to the idea of a form of transcendental unity creeping back in to an otherwise emergent understanding of nature. That "unity" could be in the form of nationalism, ethnocentrism, racism, sexism, speciesism, or any other ordering that places a single concern over that of others. It could also be a futuristic form of unity, which is found in the idea of monological and universal inherent progress. Though Haeckel challenged human/nature separation, he was unable to see how he read his own culture and biases back into nature. Similarly, though he challenged teleology in nature, he still held to a universal and monological understanding of progressive evolution (an "upward" movement) which contributed to his racist ordering of humans, and his ethnocentrism.

Haeckel was following the "ranking" model of difference that was going on in philology and early religious studies. According to this model, difference was assumed to be better/more advanced or as worse/less advanced: and everything in the world had to fit into a single history. Not surprisingly the grammar of German was ranked most advanced by nineteenth-century philologists (most of whom were German), just as monotheism was ranked as the "most advanced" religious tradition (again by those influenced by monotheism and monotheistic cultures). Haeckel did the same, unfortunately, with human beings and biological evolution.

If we can counter the tendency in human thinking to aesthetically and politically rank everything according to some monological and universal transcendent norm (e.g., the normative or ideal human or future point of arrival), then an immanent framework for understanding the evolutionary process of life opens us up toward many more possibilities for becoming that embrace more and more difference as part of the process of nature naturing. A CPR will thus hold open the future as pluriform possibilities for becoming.

Baird Callicott suggests, in his *Thinking Like a Planet*, that one way to prevent the negative side of a naturalistic worldview, such as found in the connection between deep ecology and Nazism, is to extend the boundaries of our thinking beyond nation, beyond race, beyond species, beyond religion to a planetary understanding.[49] In other words, nationalism or other identity politics, or belief in progress plus something like placing ourselves into an evolutionary and ecological understanding of the world, leads to hierarchical ranking and otherness that might be avoided were we to think our ourselves in planetary terms. The planetary contains many worlds; it is not monological or premade, but rather it is the sum of all the worlds at any given time that make up our planet. In the process of reattunement, we must always hold open spaces for future reattunements rather than close everything off into the present attunement. Before turning to my constructive end, however, I want to stick with the tragic side of Ernst Haeckel's understanding of the world and address more fully his racism, antisemitism, ethno-centrism, and nationalism.

4

THE UGLY SIDE OF COMMON GROUNDS

The general laws of the aesthetic ideal are often individualized to a characteristic particularity through a national style originating in the race, locale, religion, and the main occupations to which a nation devotes itself. The more the genius of a nation expresses itself in deeds, the more spiritual content enters into its sense of self and the more individual can become its style in art. A nation does not freely determine its destiny; it is embedded in the tremendous context of the entire life of the world, and is often limited in its existence by conditions that remain hidden to it for the longest time, which sometimes even become clear to it only in the tragic epoch of its fall.

—Karl Rosenkranz, *Aesthetics of the Ugly*

One of the ultimate consequences of placing humans within an evolutionary context is a turn toward focusing on embodiment and feeling: One might say a turn toward the whole organism, not just the rational part. In this sense, an evolutionary framework for understanding the world already contains within it a turn away from the rationalism of the Enlightenment but not necessarily away from some form of empiricism or use of reason. The sciences, which have so long been cloaked in an aura of rationality as the only way to understand the world, contain their own deconstructive turn in an evolutionary context: everything, including science, involves hermeneutics. Part of this deconstructive moment in science is realizing that human knowledge and ideas of the world are not enough to contain the world's possibilities. What was understood as good/bad, moral/immoral, right/wrong, beautiful/ugly were not as defined and universal as one might have thought. Rather, as the epigraph from the *Aesthetics of the Ugly* suggests, these values and evaluations of life are culturally, biologically, and ecologically located. At worst, the subjective nature of knowledge can be misunderstood as the universal marker by which all life is to be judged, and this universal marker then becomes the anchor for all sorts of "isms" and violence.

In a sense, this is what Haeckel and some other Romantics of his time were guilty of: They recognized that all humanity is a part of the evolution of life on the planet, but mistook their own positionality (race, culture, time, gender, nation, etc.) as the foundational marker from which to judge all other forms of life, including humans. Even though earlier Romantics such as Goethe, Herder, and Schelling didn't have a biological view of race, there was nonetheless a certain ethnocentrism to their readings of nature. Haeckel still clung to a place "above the fray" from which to anchor his value system. However, the emergence of values in an ever-evolving planetary community must acknowledge that there are worlds among and within other worlds, rather than a single world by which all things are judged.[1] Such thinking entails a new aesthetic and a new understanding of beauty. Conversely, it should also include a new understanding of tragedy and the ugly, and, these things must always be thought together as they codefine one another. As

Rosenkranz notes, "Ugliness is inseparable from the concept of beauty, since the one contains the other in its development as perpetual aberration into which it can lapse with just a modicum of too little or too much."[2] It is important to name the ugly and pay attention to the ugly, just as much as to the beautiful. We are also responsible for the ugly that is created out of and constitutes our understanding of beauty: for the ugly is part of life as well. It is not to be pushed "away," incarcerated, or incinerated but rather appreciated. The inability to acknowledge the ugly, like the inability to appreciate ambiguity, is the cause of much violence in the world. If one understands beauty as emerging from the planetary process of evolution, one must also take account of the tragic and the ugly co-created by that beauty. Otherwise, we risk backgrounding and abjecting all those things that fall out of our emergent understandings of beauty.

This chapter traces some of the tragic and negative effects that materialized in thinking about the natural world on the part of some Romantic and Romantic-influenced thinkers. Furthermore, it traces a specific instantiation of an immanent theodicy in the thought of Haeckel. Finally, it examines the tragedy of Haeckel's own thought itself, especially as it pertains to his own racism and antisemitism and how this gets used during the rise of National Socialism. If part of the implications of placing our thoughts in a planetary context entail tracing the ways in which thoughts and ideas ripple out through the planetary community, then we must face the ways in which this process negatively affects earth others as well. Because the beautiful and the ugly, good and bad are always together, no thought system or single idea is immune to the negative effects as it ripples through time and space. This is part of what it means to reattune to the worlds in which we live, from the space of a *critical* planetary romanticism.

In the end, though Haeckel understood that placing humans within an evolutionary context meant also that our ideas, religions, and cultures were placed within that context, he was incapable of developing a *critical* ecological and evolutionary understanding of thought. It was this lack of placing thought and ideas within a *critical* evolutionary and ecological context, such as the planetary provides, that led to many of the more tragic components of Haeckel's naturalistic worldview. In other words, the certainty and universality of his cultural assumptions, combined with the concept of a single world and single history, led to a ranking of diversity which placed his own culture on top, and this would have tragic consequences.

THEODICY AND HAECKEL'S RESPONSE

Haeckel knew all too well the tragic side of life. The loss of his first wife, Anna Sethe, put him into a yearlong depression that he almost didn't come out of. Throughout his life, the death of loved ones and friends, and his own bouts of depression helped secure his belief that there was no omni-god. The real cruelty of an omni-god, for Haeckel, was that He would let such terrible things happen in the world.[3] If there were such a God, for Haeckel, that God would have been worse than anything in the world.[4] And so, just as Haeckel tried to shift beauty, goodness, and truth from a transcendent realm to an immanent frame, so he sought to understand tragedy and the tragic aspects of life from within an immanent frame. However, Haeckel thought that the tragic and negative could be minimized through scientific progress. Perhaps he would find kindred spirits in the contemporary transhumanist desires to overcome death and decay altogether. This is, of course, in the end a rejection of our evolutionary and ecological contexts, which requires death and decay in order for life to continue. Regardless, for Haeckel the progressive understanding of time meant that everything could be perfected, and this meant that many of the tragedies of nature could be done away with. This idea had and still has dangerous consequences.

He understood that nature, though perfectible, was indeed imperfect. Deformities, illness, and social ills were all things that we humans must deal with, and there is no god to save us. His answer to imperfections, was in a sense to take the role of god into human hands: instead of modifying notions of perfection, or what good, and beauty, and truth might look like from an evolutionary perspective, he transplants those old, transcendent notions imperfectly onto evolution and makes humans more responsible than ever for achieving them and eliminating the negatives. Similar to Nietzsche, Haeckel gets rid of the omni-god, but replaces the agential void left behind with humans (at least some humans). As one author notes, "Haeckel based ethical judgments on the health of the species, making eugenics his ultimate ethical system."[5] He understood humans as having evolved the unique capacity for detecting beauty, goodness and truth in the world, and he went a step further in suggesting that we should (through the sciences) intervene to promote these ideals. This is not in itself a bad thing, but it becomes problematic when one mistakes their own culturally defined understandings of beauty, goodness, and truth as "natural" and universal.

In other words, though Haeckel was placing humans into an evolutionary framework, including our values, the values he read back into the natural world were merely a hierarchy turned immanent, leading to the most progressed—Good, Beautiful, and True— as his own European, scientific civilization. We could, through education (or "civilizing"), turn lesser humans and cultures into better ones. In fact, this was a human duty because evolutionary processes did not ensure progress; there could also be regress.[6] We could, he thought, through euthanasia and other genetic interventions end the misery of those severely disabled or insane.[7] Through eugenics, we could perfect the evolutionary trajectory of humans and other organisms. Haeckel was by far not the only scientist to believe these things, but the implications for such beliefs would prove to be tragic for so many humans and many other members of the planetary community.

In a similar way, the kind of transhumanism that persists today, which reifies certain aspects of humanity and certain types of ideal humans, is still at play. In her book, *Astrotopia*, Mary-Jane Rubenstein critically examines the transhuman eschatology's of Elon Musk and Jeff Bezos. Both argue in one way or another that planet earth is not the future home of humans. Musk wants to "occupy Mars," while Bezos has in mind space stations for future humans. Rubenstein writes, quoting Musk:

> "This world is coming to an end" . . . "but trust in me and I'll bring you to a new world, where you will finally be free." Free from death, free from Earth, free from gravity—at least most of it—and even free from international regulation. (Read the fine print of Musk's Starlink contracts, and you'll see him declare Mars "a free planet," over which "no Earth-based government has authority or sovereignty.")[8]

What type of "human" will be preserved in these future scenarios? Who gets to "flee," and what of humanity will be preserved? The answers to these questions reveal visions of an ideal humanity that surely will leave a lot of humans (not to mention nonhuman life) behind. Taking the future of humans out of this planet involves asking the question of what/who will continue.

Furthermore, and as Latour notes, because we are entangled with all other life on this planet—through evolution, ecology, genetics, and microbiomes, for instance—we lose what makes us human in the first place if we lose or leave the planet.[9] The transhumanist view is, from my perspective, just a new version of eugenics and escapism. We should and have to imagine different futures, but I would argue that these futures should avoid the pitfalls of eugenics and include the context from which humans and all other life emerge: viz., the planet earth.

From another angle, Haeckel's turn to dealing with theodicy parallels the constructive turns that Foucault and other postmodern thinkers would later make. There was a shift in terms of the ways our bodies, thoughts, and emotions are shaped through discipline and punishment away from the churches, and toward prisons, asylums, hospitals, and the laboratory.[10] Scientists such as Haeckel were attempting to wrest authority from religious leaders and place humans within a naturalistic framework, but failed to understand that they were importing a hierarchy of values from the theological realm and transposing it onto the natural realm. In some ways this is just an expansion of Carl Schmidt's insight that all concepts of the political variety are theological (at least in the modern Western world); but I would add along with others working on political ecology, that concepts of the natural, theological, and political are all wrapped up together.[11] Instead of focusing on the "beyond," the immanent hierarchy of value would take on the character of depth, shadow, and darkness (at the bottom), and light (at the top). One can see this in Haeckel's imposition of darkness and light onto the hierarchy of human evolution, but also in metaphors of science bringing us into the light and out of the dark ignorance of religion.[12] This is perhaps paralleled in something like Freudian psychoanalysis in which one's health depends on bringing everything from the dark depth to the bright surface. The problem is that these culturally and historically located ideas and hierarchies were read into nature. The lack of awareness of this enabled many scientists then, as today, to read their own subjectively located value systems back into the rest of the natural world as "natural." Not to mention the projection of darkness simply covers over the construction of one's own reality as "natural" and most often "good," and hides the violence and tragedy by which that positionality exists.

Though Haeckel traveled much of the world, cataloging and recording the diversity of things, peoples and places he encountered with his paintings and sketches, he was still very much housed within the framework of a nation and a place. As I argue, this attachment of a Romantic sensibility to a specific place or bioregion, can easily transform into language of purity, borders, and what counts as natural or unnatural. This is the violence of mistaking subjective (cultural-biological-ecological-historical) location with objectivity. Scientists then, as now, had for the most part not developed a sense of "strong objectivity" in which one's subject location is understood as part of what influences the interpretation of data.[13] If we are part of the rest of the evolving natural world, the idea of "strong objectivity" suggests that our own subjectivity (our location in life) must be part of any equation we make or any epistemological claims we make. There is no "neutrality" but rather neutrality masks particular culturally located habits, ideas, concepts, and affects as "universal" or "natural." This is precisely why we need multiple perspectives and why we need critical theories as part of a CPR.[14] Transhumanists today are often guilty of this same confusion: projecting a certain form of "humanity" as that worthy of saving at all costs.

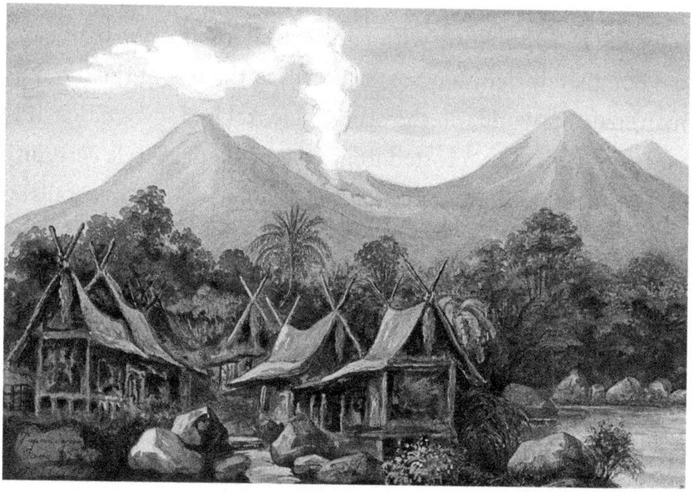

FIGURE 4.1 Ernst Haeckel, *Papandayan Volcano, Java* (1901). Perhaps a Haeckel-painted watercolor with the 1815 eruption of Mount Tambora in Indonesia in mind. This eruption changed the climate of the earth and was written about by many scientists and philosophers just before Haeckel's time. It was an event that raised issues of "theodicy" in relationship to "nature" for many. (CPA Media Pte Ltd/Alamy)

WHAT WE CAN LEARN FROM HAECKEL'S LACK OF HUMILITY

Ernst Haeckel was no apophatic, postmodern thinker. It was, in the end I argue, his leaning upon the old crutches of monotheistic-style certainty (there is but one truth), his understanding of progress, and his housing of Romanticism within nationalism rather than the planetary, which led to some of the most egregious implications of his thought. Haeckel had to walk a new tightrope that many scientists before him had not yet trod (and many still do not). Though to the postmodern mind, it is quite a "normal" situation to navigate the difference between an interpretation of nature with nature itself, on the one hand, and not completely deny any sort of access to the rest of the (nonhuman) natural world, on the other; Haeckel had few guides. It is no surprise that he resorted to mono-theological style certainty and combative arguments for his version of the naturalistic truth.

Haeckel himself had to be open to a "different" type of knowledge in that he came to see received knowledge as wrong, and that any new method for understanding the world should have built in mechanisms for allowing knowledge to be challenged and changed: He himself was constantly reattuning to the world as he studied nature in more detail. He understood that language, ideas and interpretations did not equal reality. Yet, he also understood that we must express ourselves through language, and that we could only know nature from within human language and from our various human perspectives. Similar understandings of language and reality are expressed today by many postmodern thinkers. In Mary-Jane Rubenstein's book *Strange Wonder: The Closure of Metaphysics and the Opening of Awe*, she traces the ways in which wonder continuously opens us up to different ways of being and becoming in the world through its insistence on uncertainty, which leaves open ends or a gap that remains between language/ideas and the world in itself. Metaphysics, in contrast, the language of certainty, often closes entities off from the world, and creates certain objects in the world, and the world eventually into a collection of certain objects. This closure of metaphysics is of the same type that Carolyn Merchant points out in her now classic *The Death of Nature*. The complete closure of chance and wonder in the natural world through a mechanization, which was little more than applying Newtonian mechanics to all of the natural world (geology, evolution, physiology, etc.), just brings the closure of metaphysics down to earth. If human metaphysics are human, then it turns everything in the earth into standing reserve for human beings.[15] Said another way, the science of reduction and production is based upon abstracting entities from their evolving ecological contexts and in the end creates a world in which humans are more and more abstracted from their ecological and evolving contexts. But not, of course, without great planetary costs.

Haeckel, among other Romantics, was responding directly to this type of metaphysical closure: whether it be in the form of the idealism/dualism of dogmatic theology, or the closure of a reductive materialism that he saw going on in what was, at that time, disparate and emerging specialized sciences. Haeckel's thinking in more geological, ecological and evolutionary ways (rather than more cosmological or metaphysical but not to their exclusion) required a more immanent and less hierarchical understanding of how we might know the world around us in the first place. This earthly thinking requires our thought to get dirty with the evolving planetary contexts from which they emerge. However we might know the world or even the universe / cosmos, it must be as embedded creatures within our planet, and this brings to the fore the issue of multiple possible perspectives.

The unknowing or unlearning of received knowledge and reattunement on Haeckel's part was in some ways due to his refutation of the idea that humans have access to bare facts in the world. However, even with this safeguard against hubris, Haeckel fell victim to the perils of certainty, of not recognizing his own located interpretation of nature as just that: an interpretation. The perils of not understanding history as histories when it gets placed within nature is evident in the evolutionary readings of humanity during Darwin's and Haeckel's time that led right up to Europeans being placed at the top of evolutionary trees.

Progress, as in the Enlightenment idea that the sciences (human and natural) were the one true knowledge that will make everything better, was read into the evolutionary record in part because it was not yet possible for early evolutionary thinkers such as Haeckel to not mistake their own located-ness as an objective position. Unfortunately, Haeckel's projected certainty, his love for instigating fights in his arguments against dogmatic theology, and the horrors of two world wars in the twentieth century, helped to paint him as an example of the warfare model of science and religion. Though this is one (semi-) valid perspective, I think we can reread Haeckel in a way that we learn from his mistakes and resituate him as a champion for a nonreductive religious naturalism that draws from many different humanities and sciences.

What we need now more than ever are a proliferation of readings that open us on to possibilities from our past that were "shut down" by unforeseen historical consequences. We need "multiple readings of the past" to help us navigate the detritus of history,[16] or to find "pearls" that may be used for thinking in new and different ways.[17] One such pearl is Haeckel's reading of a scientific worldview that is nonreductive. A scientific view of the world does not have to be reductive but can be one of emergent newness.[18] If time is more of a bubbling phenomenon or more rhizomatic in the Deleuzian sense, then demanding a single reading is like demanding that a person can be reduced to a single trait or character description.[19] Like people, time and history also contain multitudes. From a contextual reading of Haeckel—one that understands there is no single "science" or "religion" but only sciences continually defining themselves in dialogue with religions, folklore, and magic, and religions continually defining themselves in dialogue with other cultures/philosophies/belief systems/practices/sciences—we can read

Haeckel as developing a different type of religion from within his own context: a religiously informed naturalism. He utilized scientific observations, philosophy, theology, and many life experiences in order to construct his understanding of the world, but all of this becomes flattened by a monolithic reading of Haeckel as merely the enemy of Christian theology. So perhaps different interpretations might help to unlock different possibilities for thinking with the present and about the future.

First, we must be more diligent in recognizing that there is no nature outside of our cultural locations: there are only nature-cultures.[20] Put in the language of the philosophy of science: there are no uninterpreted facts. This does not mean, as with some postmodern thought, that only human language and ideas matter or that everything is relative, or that there is no reality outside of language. Rather, it means that we are embodied nature-cultures, bio-histories, and however we understand the world it is through these lenses, which co-construct each other. Though Haeckel understood some version of this, he was quite certain that the scientific understanding of nature would supersede all other understandings and solve all the world's riddles.

Second, unlike Haeckel who was following the general, progressive Western understanding of time during the nineteenth century, we do not have to understand time and evolution as inching toward better and better ends or goals. Teleology seals off the future and allows for certain rankings of extant life. Haeckel's interpretation of the evolutionary place of other people's vis-à-vis Europe is an instance of this. Anyone developing a type of religious naturalism, such as a CPR, should never forget that uninterpreted natural laws and scientific facts cannot replace the gap left by killing off uninterpreted revelation from God.

Third, despite our best efforts, no single person has control over the ways in which their ideas and actions will be taken up in the world. The future is open to many (not all) possibilities, and not yet determined. An uncritical use of Haeckel's monism to support a contemporary environmental ethic today runs the risk of smuggling in some of the more colonizing, racist, and antisemitic implications of Haeckel's thought. Despite Haeckel's best efforts, some of his ideas materialized in the world in negative ways that led to serious divisions and violence, from which Western environmentalism, and the general intellectual relationship between "religions" and "sciences" still suffer today. Though Haeckel threw out what he considered to be certainties based upon "religious" ideas and dogma, he replaced them with scientific certainty and even smuggled in some of the same mono-theological concepts (such as the singleness of abstract truth and progressive universal time).

THE USE AND ABUSE OF HAECKEL'S THOUGHT IN NATIONAL SOCIALISM

Haeckel has often been caricatured as a mere step along the way to Nazi medicine, if not the champion of Nazi medicine par excellance.[21] Though I (following Robert Richards) do not think it can be argued, as Stephen Gould and Daniel Gasman have argued, that Haeckel paved the way for Nazi biology, I do think there is a huge caveat to my disagreement.[22] He was a man of his time and during that time "evolutionary" theories were applied to just about everything in a value laden way. For him, the fact that the sciences he knew and loved were taking off in the European world—though he does remark on the advances of the Muslims and the Chinese in earlier times—was evidence of a superior evolution in culture via-a-vis what he encountered as more "primitive" cultures in his travels. In the same way that Christians thought of themselves as superior to those practicing other faiths, so now with those who adhered to the emerging sciences and the Enlightenment. Such an evolutionary understanding can and did play a role in the horrors of Nazi biology and in eugenics elsewhere in the world. This was also the case with other sciences—natural, physical, social, and human—in the nineteenth century.[23]

He did, as we all do, interpret nature from his own context, and allowed for too much closure of "the environment" and culture in his interpretation. Isolating European scientific thought from its context and making it the most evolved, he read back into nature his own socio-historical location—increasingly a nationalistic one after World War I—as if it were merely natural. This led to a host of Eurocentric readings of evolution, most famously captured in his drawings of the human evolutionary tree. It meant that he backgrounded or covered over all the different contributions made to the emerging scientific worldview from other cultures and times.

He, along with many others of his era, thought Europeans were at the top of the evolutionary scale. In this sense, though not in a teleological one, evolution was progressive and, for Haeckel, perfectible. I'd argue the notion of a progressive history, and one that can be perfectible, was something Haeckel uncritically smuggled in from the theological dogma he fought so vehemently against. In addition, the closure of the world in the form of his mistaking his own cultural beliefs in general as "natural" were also hangovers from the metaphysically hierarchical world he argued against. Finally, it seems that the events crystalizing nationalistic sentiments around World War I clouded Haeckel's (among others') readings of evolution in ways that play into the "imagined community" of the nation-state.[24] This ethnocentric and nationalistic reading of diversity in an

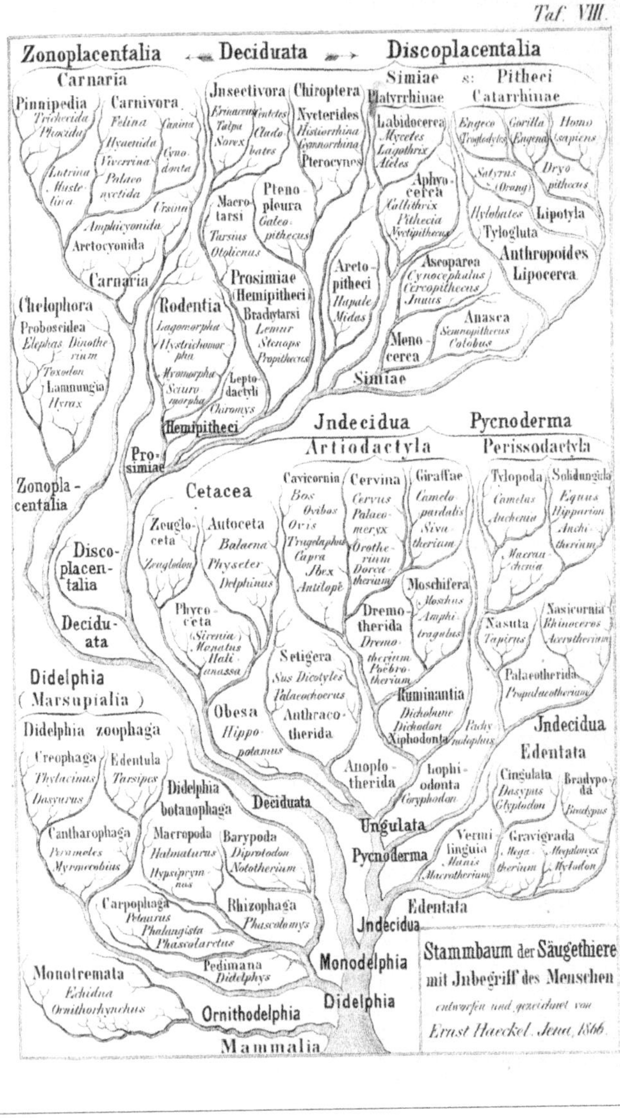

FIGURE 4.2 Ernst Haeckel, "Stem Tree of the Nine Species," in *Die Natürliche Schöpfungsgeschichte* (1868).

evolutionary perspective does seem, however, to go against some of the main ideas that come out of Haeckel's reading of evolutionary thought. Again, if placing humans in an evolutionary context meant that there were no real species distinctions and that all reality must be interpreted, one's own thoughts should be seen as just that, an interpretation. There is no inherent reason why diversity should be ranked, other than some sort of ethnocentric view being read back into the natural world. Haeckel, in this regard, was very certain of his views of nature and the place of science and European culture therein.

The outcome of epistemic certainty, belief in the perfectibility of humans, ranking of diversity with white Europeans (and even Germans) on the top, along with the rise of nationalism and the emerging science of eugenics, would have tragic and disastrous effects in the early half of the twentieth century: not just in Europe with the Nazis, but also with the sterilization of "undesirables" and other eugenic projects in the United States and elsewhere. Haeckel's failures and eventual lack of vision beyond nationalism do not have to be our own, and we must learn from these previous mistakes. Concepts of beauty, truth, and goodness, even if seemingly based in the "natural world" should always be critically scrutinized for their cultural-historical limitations. We should not fall into the phenomenological trap of thinking that we have access to unmitigated beauty, truth, goodness, wonder, or awe, which then become guides to unmediated experiences of our "true" selves, our "true" feelings, or the "true" reality of the natural world. They always include a concept of their opposites that can be used to oppress and eliminate that which is disgusting, frightening, and repulsive. The "common grounds" of common sense are always based upon the undercommons: the detritus and debris of that which has been cast out in order to build a common understanding.[25] These are socioculturally located phenomena and are never merely intuitions of nature. We need critical theories to scrutinize our own locations; we need multiple perspectives of the many worlds in which we live in order to understand how those worlds affect different bodies differently. We are never "at one" with ourselves or with nature in any sort of unmediated way. This collapse into unmediated nature aesthetics despite his warnings against "bare" interpretations of nature was, perhaps, one of Haeckel's largest mistakes.

Having said all of this, Haeckel, like all of us, contained multitudes. He was not always consistent and held many different and sometimes competing ideas throughout his life, many but not all of which he wrote down. Just as we must resist the temptation to make a clean and rosy interpretation of a historical figure in an effort to critically retrieve some of their writings and thoughts for the present, so, too, must we resist the urge to too quickly condemn a historical figure to their worst parts. Obviously, there are exceptions to this in egregious cases, but I do not think Haeckel fits into this exception. If we are too quick to cancel and condemn, we will have to throw out most thought up to and including the present, leaving us with a myopic vision of the ecological and evolutionary development of thoughts and ideas over time. Furthermore, we would be condemning ourselves to repeat the mistakes of the past.

Instead of a cancel culture, we need a strong culture of what Loretta Ross describes as "calling out" and then "calling back in," even and perhaps especially in our scholarship.[26] Cancel culture will ultimately end up creating a world of one. It ultimately destroys the pluralism it means to protect. This is not to say that some ideas shouldn't be rejected, but rather that we should work with a pluralistic community in order to create a planetary polis from which to think about the types of planetary futures we want to help co-create. Accordingly, though one can clearly identify and condemn Haeckel's racism, ethnocentrism, and antisemitism, and his pivot away from peace advocacy and toward German nationalism during World War I, there are also several reasons to resist the temptation to throw him out entirely.

He adamantly rejected any type of imperialism when it came to free thinking: the Nazi's were propagandists par excellence and opposed to free thought. Haeckel would have despised their dogma as much as he despised theological dogma. Haeckel died in 1919, and the Third Reich did not come into power until more than a decade later; it's not really clear how he could be blamed personally for the ways in which evolutionary theory in general was taken up by the Third Reich, except from an anachronistic point of view which favors a single narrative of history over the complexity that is any historical situation. In fact, his writings, along with many others, were banned by the Nazi's early on. As Robert J. Richards notes,

> Among the works to be expunged were those by "traitors," such as Albert Einstein; by "liberal democrats," such as Heinrich Mann; by "sexologists," such as Magnus Hirschfeld; by "all Jewish authors no matter what their sphere"; and by individuals advocating "the superficial scientific enlightenment of a primitive Darwinism and monism," such as Ernst Haeckel.[27]

Haeckel argued that the methods of science were superior for understanding nature to methods from religion and theology (whether Christian, Buddhist, Confucian, Islamic, or other). He thought everyone could be "educated" and made better if education was changed to allow for more direct study of nature and less study of religion. This was part and parcel to his understanding of perfectibility at the societal level: those "lesser" cultures could "improve" through the study of nature. What changes in education and culture could not improve on, eugenics could take care of: viz. (and tragically), the severely mentally and physically disabled and the insane. While eugenics itself was used in Nazi ideology and biology, Haeckel never intended to single out entire populations of people for elimination based upon their race, culture, ethnicity, or religious or other beliefs.

The individual capacity of all humans (minus, for Haeckel, the severely disabled, despite his own crippling depression) to learn and be "perfected" was, in fact, democratized by Haeckel's depiction of the theory of evolution, which the Nazi's did not like—even though Haeckel believed European cultures held the best form of education through the natural sciences. He argued against aristocracy and for a more democratized understanding of the potential of people, regardless of where and how they were born. One was not born a ruler, or superior, but education could make anyone great, and this was not something that fascism could allow. He argued, for example, "that those of royal blood would be chagrined to learn that during the first months of development, the human embryo of a noble or middle-class individual could not be distinguished from that of a dog."[28] It was his understanding of evolving nature writ large that led him to reject dogmatic positions suggesting any inherent teleology in nature or inherent superiority of one group over another. The self-understanding and propaganda of the "Third Reich" was clearly a teleological one in which certain people were inherently superior to others.

None of this should take away from the fact that Haeckel was racist, antisemitic, and ethnocentric. We must remember this when we engage his thinking. The dangers of Haeckel's belief in European superiority and certainty about what might be called "better" and "worse" housed within a sense of nationalistic identity should force us to take pause in our contemporary contexts (in which nationalism is again on the rise and claims about who/what is a part of a given national identity abound). Beyond what Richards has pointed out as good reasons for not laying the blame of Nazi biology at the feet of Haeckel, or Romanticism in general, and beyond what I have already argued here, I would like to put forth two more reasons why Haeckel's own thought might be currently used to resist something like National Socialism, and they have to do with his resistance to overspecialization and to reductivism.[29]

One of the problems of increased specialization that Haeckel saw in the natural sciences was about the alienation of one discipline from another. He wrote,

> The vast structure of science tends more and more to become a tower of Babel, in the labyrinthic passages of which few are at their ease and few any longer understand the language of other workers. In these circumstances, it seems advisable, at the commencement of our philosophic study of "the wonders of life," to form a clear idea of our task. We must carefully define the place of biology among the sciences, and the relation of its various branches to each other and to the different systems of philosophy.[30]

In many ways, Haeckel's attempt to articulate a naturalistic worldview in terms of monism was also a battle against the fragmentation and overspecialization within science, and the alienation that leads to. Monism was inherently opposed to the hyperspecialization of and separation between disciplines. Indeed, "its holistic, spiritualized understanding of science ran counter to methodological trends within the natural sciences and contradicted the functional differentiation that was assumed to govern the relationships between the disciplines."[31] It is this very fragmentation that feeds a reductive, productive model of science.

For Haeckel, it was a problem that chemists, geologists, physicists, and biologists increasingly couldn't speak one another's language. The overprofessionalization of science led to too much fragmentation within the sciences, and Haeckel called for a more integrative scientific approach. His own naturalistic approach called for much more direct observation of nature and making connections between various sciences in order to construct a nonreductive naturalistic worldview that was free from dogmatic impositions of all kinds.[32] As such, the professionalization, mechanization, and fragmentation of science under the Nazis would have been terribly problematic for Haeckel.

Though some have argued for the inherent connections between wholism, deep ecology, and Nazi ideology of "blood and soil,"[33] I, following thinkers such as Val Plumwood, argue that the fragmentation, mechanization, and professionalization of science that took place under the Nazi model was the larger problem.[34] Everything was treated as resource for the future becoming of the perfect society—people, other animals, and the rest of the natural world alike. This is the model of industrialism placed in high gear, with perhaps a veneer of place-based environmentalism added in order to help draw the great distinction between those who were counted within the circle of moral concern—i.e., those Germans who were part of the German blood and soil and not foreign to it—and those that could be counted as undesirables: Jews, mentally and physically disabled people, queer folks, intellectuals, communists, Roma, et al. Within this context, reason was placed into the service of a reductive model that understood only a select few as not "standing reserve" or eliminable for the march of Nazi progress.[35] Furthermore, as many have pointed out, the fragmentation of the engine of Nazi progress meant that most people had specialized positions in which they did not see the full reality of what was going on.[36] The study into "nature" was hitched to an ideology of progress, and all tasks therein were geared toward producing a specific outcome. While the Nazi ideology may have been a wholistic one in which everything and everyone had a proper role and place, and those that did not were meant to be cut out of the "healthy" body, their methods were that of reduction and production. This model of science would have been strongly opposed to by the likes of Haeckel.

In addition, it would make sense that any Romantic basis for science would be attacked because it went against the reductive, productive method that creates things and wealth for some people. The cold objectivity of the reductive and productive model allows for the cruelty of its technology and methods. Having a "feeling" for nature, or some sort of wonder and awe doesn't allow for transforming life into things that are useful for some human beings. I would argue that one reason Haeckel and other Romantics became linked with Nazi science after World War II has to do more with the rhetorical linkages between evolution, atheism, Marxism, and Nazism. This requires a bit more unpacking.

Just to reiterate, the sciences had not yet settled on a reductive and productive method as the universal basis for the natural sciences, even as World War II began. Wilhelm Bölsche, Gustav Fechner, Haeckel and other Romantic German scientists had been articulating methods that were everywhere from reductive materialism, to wholism, to something in between. The German Romantic scientists clashed with the British empiricists, who were largely reductive and productive in their approach to the sciences (though not all of them). In World War II the race to create the atomic bomb and develop other weapons (chemical and technological) in order to win the war led to a whole generation of scientists—chemists, biologists, and physicists especially—being sucked in to a reductive and productive model of the sciences. The Nazis developed this model with gusto. But the Western Allies, especially the United States, did as well. The technology transfer of the sciences during World War II was so successful, in fact, that after the war these methods were applied to industrializing agriculture around the world with the "green revolution" in agriculture, and after this, toward improving communication, transportation, and production technologies in general.[37] The goods of this technological advance—ending poverty, extending life, increasing quality of life—were focused on to the exclusion of the ugliness it created—pollution, displacement of peoples, climate change, and global economic inequity. The domination of this reductive and productive model was not fully challenged from within the applied sciences until Rachel Carson published *Silent Spring* in 1962—with the exceptions of some scientists who protested this model and sought one of wonder, and philosophers of science such as systems theory or those based upon process models.[38] Haeckel was also against such a reductive model of science.

His clash with theists and materialists, his views on cultural evolution, and the posthumous charges against his work as a support for Nazi ideology all provided fodder for anti-Haeckel polemics. The "creationists"-vs.-"evolutionists" battle in the United States, for instance, is in part the result of a group of Protestant ministers in Germany known as the Keplerbund organizing to combat Darwinism in response Haeckel's polemical arguments for Darwin.[39] Their arguments began to link evolutionary theory with Marxism, atheism, and then later Nazism, (all things ugly, evil, and false) as opposed to those holding onto Christianity, creation, and democracy (all things beautiful, good, and true).[40] These same types of ideas would be parroted in the so-called Cold War (communist = atheist = bad, Christian = democracy = good), and even into the post–9/11 world (predominantly Muslim nations = bad, predominantly nations based on almost any other belief system except communism, which is still bad = good). This has seeped into the Christian nationalism that opposes science and common sense at every turn, not to mention that which opposes all "others" other than white, American, Christian, straight, and cisgendered. As long as science is still in the business of making money for the "good" guys, then it is welcomed in. Once science begins to challenge the dogmas of the "good" guys, then it is also attacked as part of the woke agenda. The science surrounding sex, sexuality, gender, vaccinations, and climate change have particularly come under attack by religious nationalists of many stripes.

These politically motivated schisms between "religion" and "science" mean that many people who considered themselves "religious" were and are trained to be opposed to the scientific understanding of nature, and vice versa: many who are scientifically minded and trained tend to see only a caricature of religion as a backward way of thinking. Such divisions would eventually also contribute to religious suspicion of "environmentalism." My point in articulating some of the implications of the more polemical components of the evolution /science vs. creation/religion debate is not to let Haeckel completely off the hook. It is rather, to show that Haeckel has been read through this geopolitical and polemical lens, a lens that does not give his own historical context the fidelity it deserves.

Even if the problems of nationalism, reductivism, and productivism are more numerous and of greater magnitude than any links between Romanticism and Nazism, linking any type of worldview to nationalism is, in the end, problematic. Attaching a naturalistic worldview to a nationalistic identity will eventually lead to a ranking of diversity, and a placing of nature (including people) into the service of the nation. Just as attaching a theology to a national or racial identity will. What might happen if we loosen the boundaries of naturalism (religious or non) from that of nationalism, and begin to understand some version of nonreductive naturalism, or critical romanticism, from a planetary perspective?[41] What, in other words, if we loosen the technologies and habits of affect from "the nation," from "us and them" mentalities that make the other into one's opposite, and began to think in terms of planetary affects, creating what I am calling a CPR for the earth. From this perspective, one must pay deep attention to planetary others and the undercommons that are made from common grounds. One must see both the beautiful and the ugly effects of one's thoughts and actions, and how different bodies are affected differently. The ugly cannot be cut off from the beautiful, but both make up the planetary community and depend on one another.

HAECKEL IN AN ERA OF GLOBAL CLIMATE CHANGE, GLOBALIZATION, AND PLANETARITY

It is important to both read Haeckel in his own context and learn from his tragic mistakes. What does it mean that Haeckel's whole understanding of "nature" was developed before any sort of modern environmental movement? It developed before a time in which nature needs saving, so to speak. What does it mean there was nothing like the globalized world of climate change we experience today? Even though the Romantics were reacting to the Industrial Revolution, and there was some talk of climate change and environmental problems, there was not yet a widespread sense and public conversation that human forcings could fundamentally change the environment in irreversibly negative ways, much less the climate. This bit of context deserves proper exegesis as part of a "geostory" just as much as any historical contextualization of religious scripture does.[42] Globalization and global climate change have forever changed the way that we understand "nature" and our place therein.

In terms of the ugly or tragic side of Haeckel's own thought, most of the negative outcomes of his thought were related to his nationalism and to his holding on to a specific version of progress. Even though he traveled the world, and even though he sought to unify all of life into a single, evolutionary perspective, he ranked that life hierarchically according to his own location and projected his own historical-social location as "natural." He operated under what we might call a nationalistic romanticism. He knew there were no bare facts, but he could not extend that criticism to his own (European and German) location (nationalistic romanticism). Where foundations and a priori truths stand, critical thinking ends. From where he stood, he was unable to see the ugliness his nationalistic aesthetic created; rather, he focused on how to perfect or beautify the world according to that nationalism.

The ranking of life from such a perspective misses the fact that we are always and already mixed up historically, biologically, and culturally, and that diversity and difference do not necessarily have to be ranked. Furthermore, it disables the critical investigation of what one thinks of as aesthetically beautiful/ugly and morally good/bad. This is something we might be able to see more clearly from a globalized, critical planetary romanticism. Yet it must still be asked, Why was someone like Haeckel such an iconoclast on so many issues, but not on these racist and ethnocentric assumptions? I suspect it may have something to do with the deep affects and desires of an era before widespread globalization and global perspectives. In other words, our emotions, bodies, habits, desires, feelings, hopes, and dreams no matter how "far out" can never fully move beyond the histories and realities that shape them. We are not merely rational creatures but affective creatures as well, and our reason is housed in these affects; it does not dangle alone in the wind free of context, and indeed the very contours of reason are shaped by certain affect filled contexts. As Donovan Schaefer argues, "*Feeling is necessary for thinking; there is no thinking that is not feeling*."[43] This is evidenced by the contemporary reaction of many around the planet to pluralism and globalization in the form of some return to an imagined nationalism and localism.

In addition to our globalized planet, the context of global climate change has forever changed the way we might understand human agency in relationship to the rest of the natural world. It was easier for someone before the twentieth century to argue that humans might be responsible for the "perfection" of nature through eugenics, landscape management, and education because the full effects of the fossil-fueled Industrial Revolution had not yet become obvious. That human agency is wrapped up with the agency of other organisms and systems on the planet in ways that make most of our planetary problems "wicked" was something that most would not have understood from a pre-twentieth-century location.[44] Though Haeckel was nonreductive in his understanding of the natural world (and humans therein), he was still incapable of freeing his ideas of humans-in-nature from the confines of progressive chronological time, which was housed within aesthetic and narrative frames that were increasingly nationalistic.

These frames along with a reductive and productive understanding of the natural sciences helped to produce and were reinforced by nation-states. This produced an unwarranted amount of certainty, which one finds in Haeckel's belief and confidence that every riddle of the world would eventually be solved through advances in the sciences.[45] This "shrinking" of scope mirrored also the "shrinking" of the sciences to atoms (in physics) and a bit later genes (in biology) as fundamental units of reality that might lead toward the successful recreation of nature according to human will and desire. Such intoxicating lures in terms of human control ought to be guarded against, if for no other reason than that it does not map on to the reality of the messy planet we inhabit and thus ends up causing violence to many earth bodies. I argue that the ugliness the project of modern, Western progress creates outweighs the beauty it might provide for some (humans). Whereas the technologies of modern Western thought might be useful for a way of knowing that requires a smooth objective space of removal in order to make truth claims in the world, a fleshy, "sticky" way of knowing, which the placing of humans within an evolutionary perspective requires, needs its own technologies that better enable us to feel our interconnectedness with multiple earth others.[46]

EARTHLY EROTICS

"Ero-sophia," rather than merely "philo-sophia," may be what is needed here.[47] For too long in many Western traditions eros has been locked away in the bedroom or placed into the confines of individual subjectivities interacting with one another; it is now indeed demanding to "come out." Of course, I am not only talking about sex here, but the erotic desire of our embodiments that connect us to other living things, past, present and future. What might be a way of knowing and being and doing within this sticky, messy, wicked planet; how might we begin to see the ugly and beautiful as part of every human construction? Here I can only offer some beginning reflections on counter-technologies that might reimagine and reconnect humans with the evolving planetary community: thinking with the so-called new materialisms, living at a pace of uncertainty rather than progress, and understanding agency from a perspective of multiple causality.

The same type of shift reflected in Haeckel's shift toward a more immanent way of thinking, one which is critical of teleology and design, is also reflected in what many are now calling the new materialisms. The new materialisms, and I'm thinking here of people such as Karen Barad, William Connolly, and Jane Bennett, are deeply influenced by (among other things) pantheism, process thought, early emergence theory, and the work of French philosophers Gilles Delueze and Felix Guattari.[48] In some cases these new materialisms have also drawn from traditions that reflect a more immanent ontology such as Buddhism. The work of Freya Matthews with Daoism and Val Plumwood with an Indigenous-informed animism also come to mind.[49] The point here is that a shift toward an immanent framework and against dualisms (without slipping into reductive materialism, idealism, or some form of monism),[50] means that agency and desire are distributed throughout the world—in various types of entities and processes: human, animal, plant, mineral, and machine. Agency and desires are shaped by quantum and genetic habits of becoming (Barad), through technologies such as the electronic grid, the internet, and cellular communication systems (Bennett), and through the multiple forms of embodiment we come into contact with and experience as distinct on a daily basis. Histories of biologies and natures help to shape fleshy becoming into specific forms in any given moment: all functioning well below our rational, conscious decision-making brain.[51] Such an account of onto-epistemology (as we are thinking just as much as being thought) means that our analysis of any given object, event, process, or situation is inherently complex or "wicked" (depending on the terms your own discipline might use).

The constant shifts in our enfleshments are shaped by and shape other earth bodies in the ongoing process of planetary change, and causality within this changing context is not merely efficient. Efficient causality is much too mechanical in such a complex world of becoming: rather, we must look at multiple causes, and multiple agents, that are diffused over time and space. The point of such an examination is precisely not to "go away" (or as Latour suggests, not to "freeze-frame" reality) in ideas of teleology, or in ideas of what is Natural or God-Given.[52] From these spaces of removal, we begin to force the world into becoming in just those ways we name as "objective." Agency, eros, and desire are immediately marked as appropriate, from these spaces of removal, when they move the world toward the inertia of their space.[53] Certainty means becoming in just the ways dictated by these spaces of removal. And an efficient causality seems to be the quickest way of thinking about moving into the direction of said spaces of removal. The problem is that life is too messy for any human-imposed objectives, and the worlds created by such spaces eventually come undone. Too much focus on ordering in whatever you believe to be truth, beauty, and goodness without seeing the negative things created in their shadows and taking them into account, leads to violence. This was one of Horkheimer and Adorno's most important insights in *Dialectic of Enlightenment*: viz., forcing the world into reason and making the world conform to human rationality will lead to the world's unraveling. What is needed, then, rather than mere efficient causality and progress toward some sort of objective space of removal, is knowledge production that recognizes uncertainty and ambiguity and stays within a framework of what Ursula Goodenough calls a "horizontal transcendence" that doesn't take us away from our contexts, but enables us to imagine possibilities for a "not yet" future.[54] Here at the edge of this chapter, then, I suggest how an ero-sophia or a wisdom in tune with the erotic and the affects (not just reason alone) might help us stay tuned into the world in ways that are missed through knowledge based on objective spaces of removal.

EROTICS AND THE ART OF FAILURE

In many ways, Haeckel was practicing a form of "failure" that Jack Halberstam (much later) speaks of in *The Queer Art of Failure.*[55] For Halberstam, in a system that creates so much violence and tragedy for so many bodies in the name of the very ideals of progress it pushes bodies to perform in the world, what we need to do is fail. We need to act in ways that fail to perform according to prescribed norms, roles, and ideals. This will monkey-wrench the system, so to speak, in ways that the system begins to break down. For Haeckel, this meant refusing to perform in ways that supported the social, economic, philosophical, and scientific understandings of a creator God and the forms created by that god at the end of the nineteenth century. This failure is his refusal to ignore the abjected bodies of a given systemization of the world, and a call toward planetary transformation. Indeed, many luminaries must fail the orders in which they live to bring about changes. Darwin's *Origin of Species* that Haeckel so loved and imbibed was a long argument against the very idea of static species existing in the world: rather, species were instead conceptual tools used by scientists and were in constant flux, even humans. For these transgressions, folks like Haeckel, Hirschfeld, and Darwin were (and still are) criticized roundly and compared to nonhuman (irrational) animals. Part of the early critiques of them included especially critiques of their humanity, calling them apes or beasts in ways that sought to discredit their views of the world.[56]

In these cases of failure, there is some amount of ero-sophia at play. The ability to follow life outside of its prescribed roles—which are always separating species, sexes, sexualities, humans and technologies, for example—is precisely the ability to reattune to the flows of energy of life (of eros) in ways that do not conform merely to roles laid out by histories, cultures, and societies. It requires the ability to take on the abject and allow the abject to take you along to new ways of being in the world. It is the ability to discover new ways of loving and being in failure and loss: new ways that shift the world ever more to include and seek justice and compassion for the bodies that have been abjected by religious, economic, philosophical and other apparatuses geared toward some version of progress and truth or another. These abjected bodies do indeed matter, and it is with these abjected bodies that new ways of planetary becoming might be found.[57] To begin living into different planetary futures, we might begin thinking with our bodies and their intersectionality and interrelations with all of the other sticky earth bodies; we might begin thinking erotically and in planetary (per)versions.[58] The final two chapters of this book begin to imagine how a CPR for the earth might help us do this in ways that fail ideas of progress that lead to a deteriorated planet, and promote alternatives for the flourishing of our planetary future.

III

RETHINKING GOODNESS (PERSPECTIVALISM)

5

REATTUNEMENTS TO THE PLANETARY COMMUNITY

Many philosophers and religious studies scholars have begun to imagine what a transition toward a planetary understanding of the world might look like.[1] Mary Evelyn Tucker puts it this way in *Worldly Wonder*: "Indeed, the environmental crisis calls the religions of the world to respond by finding their voice within the larger Earth community. In so doing, the religions are now entering their ecological phase and finding their planetary expression."[2] Haeckel, Humboldt, Herder, Goethe, and other Romantic empiricists of the eighteenth and nineteenth centuries were indeed in the process of thinking through a naturalistic worldview based upon the natural sciences that included history, language, religion, and philosophy. We might look to these earlier thinkers as articulating what Bruno Latour has called a "geo-story" based upon the best sensory data of their era.[3] In other words, they were articulating a planetary story (geo-story) into which histories (human stories) might fit. This is, then, a meaning-making project: a *religare/relegare* binding together of information from many different sources into a meaningful whole. Now, we might also add that this should be an emergent and open whole that pays deep attention to the undercommons and abjected bodies of any given constructed meaning.[4] In processes of reattuning to the bodies that make up the changing planetary community at any given moment, one should be aware of the ways in which bodies experience a given world or worlds for better and worse. Paying attention to the undercommons, or precisely how all common grounds are made up of silenced bodies and processes, might help us to find alternative grounds for building possible new worlds, and help us keep our hearts, minds and bodies focused on justice for all planetary bodies.

Though Haeckel did not think about justice for all bodies much less the undercommons, there were ethical implications of his thought, the first of which had to do with humans being embedded in a living process as part of the earth. His articulation of a monistic worldview insisted on the poesies and plasticity of living and nonliving organisms together: which was again, neither found in the reduction to the material, nor to the ideal, but the opening of a "third" space of interplay between these two. Another third space can also be seen in the turns to the present and to embodiment in the twentieth century: reality is neither located in the past nor fully in the future, but in the emergent present where time comes together in the interaction of bodies (and in which the relationship between organisms and their environments push the world into different ways of becoming). I would argue the focus on the interaction of bodies in the present (which always includes concerns for the past and future) is at the heart of modern ecology.

In art, particularly music/painting/sculpture, still another third way is implied in understanding how ideas and materials come together in the emergent "production" of a given object, and how these objects continue to live as various peoples consume and interpret them. In literature, a third way manifests in the form of understanding texts as living, where the reader becomes a located interpreter bringing the text into a living present, and breaking open orthodoxy to the polydoxy of multiple interpretations. These are all examples of the productivity and possibilities unleashed in these "third spaces," "eco-tones," "interstices," and "borderlands," to use some of the neologisms for places in which these hybrid and evolving interactions happen. And these spaces depend to some degree on focusing on the present.

I'm not suggesting that Haeckel was arguing for all of this, but the focus on immanent contexts and interpretations that one finds in discussions of postmodernity and pluralism depends upon an understanding of histories as part of and coming from the rest of the evolving natural world. Haeckel and these other Romantic scientists of his era helped to lay the grounds for these postmodern developments. Indeed, for many cultures and religious systems that focus on immanent structures, what is much more common is polydoxy, and multiple truths, and stories. In retrieving earlier Romantic thought for what I am calling a CPR, I think it is important to highlight four major shifts that an evolutionary, naturalistic, monistic worldview helped set in motion: the death of God or immanent spiritualties; power and planetary democracy; the idea of subjectivity as internalizations of externalities; and religious imagination as the projection of possible futures. It is in (re)attuning in these four areas more than any other, that I think a CPR can be articulated for our contemporary, earthly contexts.

REATTUNEMENT 1: THE DEATH OF GOD AND IMMANENT SPIRITUALTIES

Though Nietzsche and Haeckel did not have any known correspondence with one another, they of course appreciated each other's writings.[5] They were both interested in solving the "world's riddles" through an immanent frame of reference—that is, without the need for a supernatural God. Whereas Nietzsche infamously proclaimed, "God is dead," Haeckel got to a naturalistic worldview without god through his placement of human beings within an evolutionary framework and making Nature god.[6] Both were deeply influenced by the German Romantics, and in many ways they arrive at similar immanent frames of reference through their different methods: Nietzsche through philosophy (and some biology and evolutionary thinking) and Haeckel through biology (but also art and philosophy). What I argue here is that the biological and evolutionary placement of human beings within the evolutionary framework is an important step toward Nietzsche's declarations. Without the evolutionary preparation of these early grounds, "the death of God" or "death of transcendence" would have no traction in the world. They mutually support one another.

By placing values, concepts of beauty, beliefs, and ideas, within an evolutionary framework, whatever god or ultimate reality might mean must first and foremost be about understanding the human in its relationship to other humans and the rest of the natural world. That is, theology becomes anthropology and ecology. Unlike with Nietzsche, for us, the death of God need not mean the elevation of the individual par excellance;[7] rather, it could mean the elevation of the process of nature naturing over that of any individual perspective, and even over the Anthropos as a whole. For earlier Romantics, as for new materialists, emergentists, and other modern Western thinkers who think nature and culture, and mind and matter back together within an immanent frame, agency and value are not isolated within the individual or the human but are dispersed throughout the process of nature naturing.[8] In this sense, the agency that was reserved for the omni-god, is made immanent, and then not contained by that god's image on earth (as with Nietzsche) but dispersed throughout the planetary community. Making nature (the ongoing process of life) into god means that human bodies are not the only agential, valuable or affective bodies in the world. Such a dispersal of agency, feeling, and value, a radical incarnation, if you will, must mean that human agency and affect are demoted or deflated and that the agency and affect of the rest of the natural world is elevated. We can see this in the second shift I discuss here, namely that of dependency and subjectivity.

REATTUNEMENT 2: SCHLEIERMACHER, DEPENDENCY AND DEMOCRACY

In Haeckel's shift toward understanding humans as primarily dependent upon Nature rather than God, he also drew from Schleiermacher a sort of radical understanding of democracy when it comes to knowing "nature." Rather than experts in narrow silos dictating to all others what nature is and how we fit into it, Haeckel emphasized that everyone should learn about nature themselves through studies of and experiences in nature. This was why he pushed to get more scientific education in schools and less theology, and why he encouraged direct experiences of nature. It wasn't only the scientific, rational study of nature that was important, but also the feelings of wonder and awe produced by such study and experiences in the natural world. This was part of the reason he spent months at a time traveling around cataloging and painting the natural world. These were all things that ideally everyone could and should do, even if in a more localized way.

Such planetary knowledge helps us to realize that all bodies, processes, and events are intractable parts of the overall living processes that they make up. Whatever common grounds of experience one group of people might agree on as constituting their world, they are always connected to other worlds that might be constituted by quite different experiences. What is thought of as good/bad, beautiful/ugly, true/false in one world, may not be the same in other worlds. Thus, we should begin to recognize the ways in which our own worlds and reality affect other worlds, and vice versa. We are dependent upon and interdependent with these ever-changing worlds that at any given moment make up the planetary community, and we might live in better ways if we continually reattune to the planetary effects of our own world-making.

When we put humans into an evolutionary framework, making the basic human condition one of being "dependent" upon the evolutionary process and the genetic, geographic, climatic, and ecological gamble into which each of us is "thrown," we can begin to have a self-understanding that is not tied to nationalism or neoliberalism. We can understand ourselves not as isolated individuals but rather as processes of recycled energy, material, and even "feeling" and emotion, which endure as an organized entity on this planet for a time and then return to become part of other processes and entities. From this perspective, we might find a way out of the twin problems of neoliberal globalization as usual—dependent upon the individual as an isolated economic unit—and nationalism (whether capitalist or socialist)—depending upon isolating some individuals from others.

Neither neoliberal globalization nor nationalism (capitalist or socialist) make sense in a world in which we are radically part of an ongoing planetary process. It is impossible to separate out one from the other: one person, one group of people, one species, etc. Doing so, in fact, cuts an entity off from the source of ongoing life and thus it begins to reify, putrefy, and die. Furthermore, because of the "democratization" of knowledge that occurs once humans are part of the evolutionary process (remember, there is no outside, no objective smooth space, so all knowledge is perspectival knowledge), making an epistemic cut means that we cut ourselves off from the ground of multiple perspectives. In a planetary community, cutting out multiple perspectives is only done at the cost of planetary peril. This is not, however, to say that "anything goes" or that all facts and ideas are equal. Far from it. The idea is that we forge better planetary common grounds through engaging more of the perspectives which help co-create those grounds. Indeed, it is the perspective from nowhere, that bugaboo of the panopticon, patriarchy, and monotheism with its omni-god, that should be challenged. From these disembodied perspectives, co-creating, communing, collaborating, compassion, and coexistence are all thwarted for the power-filled ideal of creating everything ex nihilo.[9] This is why immanent thinking is so threatening: it democratizes power, values, meaning, aesthetics, and knowledge, and upsets many of the hierarchies and divisions that much of dominant Western thinking has built itself on. As Mary-Jane Rubenstein has argued,

> The problem, it seems, is that pantheism not only unsettles, and not only entangles, but demolishes the raced and gendered ontic distinctions that Western metaphysics (with some crucial exceptions) insists on drawing between activity and passivity, spirit and matter, and animacy and inanimacy—distinctions that are rooted theologically in the Greco-Roman-Abrahamic distinctions between creator and created, or God and the world.[10]

From this position of planetary dependence, of planetary democracy, then, we might agree with Catherine Keller that perhaps certainty has created more violence than uncertainty ever will.[11] In place of certainty, we opt for listening, dialogue and conversation: all of which lead to the inevitable possibility that we could be converted and changed through constant reattunement to evolving planetary bodies. This type of collaborative planetary epistemology also means that we side with Foucault in arguing that all knowledge is political, but perhaps with an Arendtian twist.[12] It is not just that knowledge is political, but that knowledge is co-created in public spaces, in the polis, and from a planetary perspective this always already includes the rest of the natural world.[13] The polis is not just that which enables the co-creation of knowledge, but it, too, is created by multiple voices thinking, feeling, and acting together (even if more as a cacophony than a symphony). The planetary polis is co-created by planetary entities, both seen and unseen, and these common and undercommon grounds help to co-create planetary entities.

Decolonial and critical race scholars and theorists have called for a type of "undisciplining" of the university for similar reasons that I am calling for built in uncertainty at the edges of what we think we know. Sylvia Wynter, for instance, argues that if the "human" of the humanities is one that is the ideal, Euro-Western, patriarchal, Christian, hetero, male subject, then the very foundation of Western-style education is colonial. In addition, in the Western split of the "humanities" and "nature" (human sciences and natural sciences), the Anthropos is disconnected from the rest of the natural world and what counts as "human" becomes a huge question.[14] For this reason scholar Aph Ko argues that in order to undo all other "isms", we must first undo that great ism, anthropocentrism.[15] Carol Wayne White argues similarly that in order to really achieve humanity, we must embrace our animality and entanglement within an evolving planetary community.[16] Finally Indigenous studies scholars have long pointed out that the assumption of human exceptionalism from the rest of an instrumentalized "natural" world has long been a catalyst for the genocide and colonization of native peoples.[17] Such rethinking of Western academic disciplines also calls for a third reattunement: subjectivity as an internalization of externalities.

REATTUNEMENT 3: SUBJECTIVITY AS INTERNALIZATIONS OF EXTERNALITIES (INSIDE OUT, OUTSIDE IN)

As has already been mentioned, we can begin to think about mind and brain, culture and nature, spirit and body, history and biology together after placing humans into an evolutionary context. The placing of humans into an evolutionary context means that whatever we call consciousness, mind, spirit, culture, religion, and thought, they emerge out of the evolutionary process rather coming to the material realm from some sort of transcendent space. All of these things coevolve and coemerge together, as Donna Haraway argues in her recent book (and as have other Romantics suggested before her); they "symbio-emerge" together.[18] One of the things I would like to argue about such a symbiotic emergence is that it also means our own identities are an internalization of our surroundings: or that the outside and inside are intimately related. The seen and unseen are co-creative; the undercommons co-create the commons. Our very subjectivities are the result of reattuning to other agents, actants, and entities that are around us.

If we move beyond essentialism in identity formation, it means that our own individual uniqueness does not transcend the world but is rather of the world from which we emerge. We are a unique combination of the processes that our bodies have been shaped by, and our bodies shape those processes. It literally takes a planet to create each and every one of us.[19] Ernst Haeckel also argued for this way of thinking when he argued that whatever "interiority" is, it is an interiorization of the world around us along with the emergence of a locus of control by which we choose (to some degree, or what Ursula Goodenough calls "choosy-ness"[20]) to move in this direction or that, become this way or that.[21] In addition, he argues that entities and processes are collectives rather than individuals. As Eric Paul Jacobsen argues,

> Haeckel observed that many organisms only *appear* to be independent individuals, but can also be regarded as the specialized organs of larger organisms. From these observations Haeckel hypothesized that a simple organic individual may combine with others to form a more complex individual, or an individual of a *morphologically or physiologically higher order.*[22]

In this sense he might find today kindred spirits in emergence theory and new materialisms in articulating the idea that new levels of reality emerge from one another and then create a level of reality that is then equally as real as the one "below" it.[23] Unlike some emergence theorists, he believed that "nothing comes from nothing" so that there must be some sort of "consciousness" or "feeling" all the way down, even at the inorganic/abiotic levels.

Whichever starting point we might take, whether "something else can come from nothing but," or whether "nothing comes from nothing" and thus there must be a precursor to something like consciousness "all the way down," may not make much of a difference for ethics and politics. But what it does mean, is that the "material" and "spiritual" world are always and already together: even in the most basic elements and atoms of the rest of the natural world. Haeckel's trinitarian/triune monism here suggests a plurality of relations as going "all the way down," in the way that some theologians such as Ivone Gebarra have understood the trinitarian structure of god as the relational matrix out of which everything emerges.[24]

This type of relationality is also similar to what Karen Barad suggests in terms of "agential realism," or in terms of what Judith Butler and others argue about the abjections of identity being what defines identity in the first place.[25] For Barad, there is some sort of "choice" all the way down so that performativity becomes important even at the subatomic levels and agency becomes something that is always and already shared across "levels" of life: the quantum, atomic, physical, chemical, biological, ecological, psychological, cosmological, etc. In terms of identity formation in the thought of Judith Butler, the abject is what defines the boundary of identity: we know what we are by what we are not (the commons are what they are because of the undercommons). But those very abjections return to haunt the edges and boundaries of our identity (and other constructions) until they force identities and structures open to change toward something new. Such an understanding of identity is completely immanent, without the need for some essence or telos/form prior to or outside of relation. In other words, there is no "ding an sich" if this means something untouched by relations. The "ding an sich," if there is one at all, is the unique set of relations that make up any given entity (or nested entities within entities, "all the way up" and "all the way down"). This is an evolutionary and emergent understanding of identity. Such an understanding of identity and subject formation should also be met with an equally open understanding of consciousness, ideas, and thought.

As Hannah Arendt argues, the inner "I" is always and already plural. The inner conversations we have going on in our heads, that thing which we might call "me," "myself" and "I", is already a conversation that has been internalized from many past and present voices.[26] Our language, the ideas in our heads, the cultures we swim in, all come to us from internalizing external histories, conversations, cultures, and biologies that have been passed down generation after generation. This is why the "polis" is so important to Arendt's work, and also why knowledge of the natural world should be public and participatory. The fact that I can put Arendt and Haeckel here together, a Jewish philosopher and an antisemite, is exactly a testament to planetary coexistence: nothing and no one is off the table in this co-construction. Yes, this co-construction must be critical, as I have been throughout this book about Haeckel's racism, ableism, antisemitism, and nationalism: a CPR demands the voices of critical theories and that we belong to a shifting, common planet that we co-create with one another. We can't simply cancel or ignore certain voices that we disagree with: those voices still enter into our internal and external conversations and shape the worlds in which we live, and we ignore them or try to silence them at our own peril. This doesn't mean that we agree with them, nor that we must listen to or allow hateful speech: sometimes we have to protect ourselves first. Boundaries and constraints are important parts of planetary evolution that allow us to be different types of entities to begin with.[27] But this doesn't mean we can shy away from hard conversations, or secure hermetically sealed boundaries. One must be part of conversations with others in order to participate in the creation of the shared fabrics that make up our interiority; and one must recognize those things which are left out or backgrounded (because they still affect us). This shared fabric, seen and unseen, which everyone interiorizes in their own way, is also the space from which new possibilities for future becoming emerge. The fourth reattunement I want to address from an evolutionary perspective has to do with the projection of different possible worlds.

REATTUNEMENT 4: FEUERBACH AND PROJECTION, OR DIFFERENT POSSIBLE WORLDS

As I have mentioned elsewhere in this book, Haeckel was very much influenced by Feuerbach's understanding of theology as projection. Feuerbach's (and Nietzsche's) thought were both shaped by the long history of negative theology and iconoclastic trends in theology as well: anything we can imagine or name as god must not be God.[28] For some this line of thinking ultimately leads to atheism. However, this is not the only possibility if we look at Feuerbach (et al.) from within an evolutionary perspective. Let me explain further.

As I have argued elsewhere, it is high time that Feuerbach be reread not as an atheist, but rather as an agnostic from within an evolutionary context.[29] He is not a fundamentalist atheist in the vein of Richard Dawkins, but much more of what I would like to describe as a viable agnostic. He is part of the shift toward immanence in modern Western thought that is taking place in the nineteenth and early twentieth centuries. If we are indeed part of an evolutionary framework, and if the shifts above hold, then we are part of an evolving, emerging process and our imaginations and ideas are collective projections in this process that take shape in the world, or that matter the world around us. This means that our collective imaginings together, including our religious ones, are like various communal castings of a rod and reel: thrown out, so to speak, in hopes of catching something of sustenance. Some of these imaginings are successful in a big way, and others are not: regardless, we must reel them back in, rethink them, and cast them out again. Some catches can even be poisonous and deadly. In a world where ideas and matter are thought together, this is not atheism but the type of viable agnosticism that acknowledges the co-construction of our ideal-material worlds in an open-ended process. Ideas, dreams, hopes, desires: all of these things matter to and in the worlds around us.

I think we can begin to take on these four reattunements, resulting from the placement of humans fully within an evolutionary framework, and start casting out some planetary imaginaries. We can learn from the past attempts to do so by not falling into the traps of certainty, which close off the ever-receding edge of the becoming processes of reality. Here I focus on the emergence of one such collective imaginary that I think could be useful for thinking toward a future in which the flourishing of the planetary community is better supported.

THE EMERGING PLANETARY IMAGINARY

Perhaps for the first time in geohistory, we humans can begin to think of ourselves fully as evolutionary creatures and fully as part of a planetary community rather than first and foremost in terms of nationality or religious identity or even as an exceptional species. As Bruno Latour argues, "The nineteenth century was the age of the social question; the twenty-first is the age of the new geo-social question."[30] If anything, the result of the processes of globalization and climate weirding have one silver lining: the realization that we are and always have been intertwined with life on the planet, for better and worse. Perhaps this is also what the recent pandemic has made clear: We must act more and more as a planetary body. Interconnectedness is not always a beautiful thing; it is also the cause of much tragedy and death. We are through and through biologically and culturally hybrid. As such, we should begin to develop new imaginaries for such an "imagined community."[31] This realization moves us beyond the linear line that moves from modernity to postmodernity, or from colonial to post-/decolonial, in that we are also opened on to the more-than-human, planetary community, which has its own times, not to mention the past which is pregnant with unrealized possibilities. What we need then, I argue, is an embrace of some form of CPR.[32] Indeed, such a critical romanticism may be necessary in order to create viable planetary imaginaries.

The apocalyptic pace at which climate change is happening is due for a head-on collision with the glacial pace of governmental and institutional change. Is the metaphor of "glacial pace" even appropriate since these objects that were once known for their slow creeping millennial pace, are now disappearing at the fossil-fueled fast pace of Western modernity? Do seasonal metaphors such as "fall" and "winter" even mean anything in a world whose climates are changing? Not that those seasonal metaphors meant much to those in equatorial or polar regions to begin with; but nevertheless the "regular" patterns of weather no longer make for sturdy, recognizable metaphors. It is not just that false sense of secure national identities and boundaries (not to mention religious, ethnic, and species identities and boundaries) are being whisked away, but also our language that coevolved with a relatively stable climate that enabled the emergence of agrarian and later industrial civilizations over the past ten thousand years. Language interpolates us and binds us not just with one another but within the planetary community also: just as our bodies are braided in genetic, ecological, mineral, plant, animal and elemental flows, so our ideas, ideals, values, hopes and dreams are braided through feelings and language together with the planetary community. The processes of globalization and climate weirding are leading to untanglings and rebraidings of our past ways of being with the planetary community, and we must begin to speak and become in new planetary entanglements. In this way we must reattune to the worlds in which we live, all of which together at any given time make up the evolving planetary community.

Of course, we could just join the (at-times) deafeningly loud chorus shouting, "Eat, drink, and be merry, for tomorrow we die!" and sometimes in the face of such planetary despair, that can be a genuine bonding experience that brings peoples, places, and things together in healthy ways. I know I am guilty of it. But we should probably also begin to act and think anew per these new attunements and entanglements. What, we might ask, given the small time, energy, and agency we each have on this planet (about eighty loops around the sun if you're lucky), should we focus our energies and attention on? Obviously, the answers to this question will be different according to context. My own answers keep coming back to thinking about how things such as language, meaning-making practices, values, and concepts of time materialize bodies differently in the world. These are things I think are important to ponder even in the face of voices that might say: We don't have time for such impractical things.

If we want to address the violence inherent in modern Western thought, then we might need to undiscipline our thinking.

How might we begin to reattune to our evolving, embodied contexts on a planet marked by climate change and globalization? What types of planetary futures might we coimagine and begin to work toward? These questions, to my mind, are at the heart of understanding human conditions in the twenty-first century and exploring them will be important for co-constructing worlds that work toward the flourishing of the planetary community. Unfortunately, without projects such as this, peoples, governments, institutions, and infrastructures seem to be falling into five primary reactions in the face of climate change and globalization.[33]

The first is a return to nationalisms and parochialisms, adopted by many people on the so-called conservative and right end of the spectrum. Circle the wagons, retreat from global connections, and take care of one's own citizenry and people. Of course, these types of nationalisms usually (if not always) lead to discourses and actions that are ethnocentric, xenophobic, and racist. The question of who counts as a citizen of X country often ends in answers that lead to the purification of a country's identity. I'm thinking of the return to nationalisms across Europe and discourse around Muslims and Syrian refugees; the return to nationalism in the United States with Trumpism and the backlash against immigrants, Black Lives Matter movements, and anti-wokeness discourse in general; Hindu nationalism in India that seeks to exclude Muslims from full citizenship; the neo-Confucian based nationalism in China and the treatment of the Uyghurs there; the attempt by conversative Muslims in Indonesia to equate Indonesian identity with a particular brand of Islam; the situations between Russia and the Ukraine, and between Israel and Palestine. The list goes on. It seems we haven't learned from the hard lessons of two world wars.

The second group follows a sort of localism or bioregionalism in the form of a return to "place" and a return to more "natural" ways of living. These ideas can be found among those on the far political left. The answer to the problems of climate change and globalization is to retreat to local living. This, of course, is not possible for people who can't afford to buy land and exit the global economy (not that anyone ever fully can). In addition, these movements often fail to take into account the history of Indigenous peoples whose lands these movements are built upon. Interestingly, the far left and the far right make strange bedfellows when it comes to things like the anti-vax movement and promoting policies against globalization.

The third general reaction is the continuation of the globalization of neoliberal economics and policies as usual, with social justice and environmental concerns tacked on. This is usually adopted by the so-called liberals who are on the center or center left of the political spectrum. Indeed, it is this very neoliberalism that has fed climate catastrophe and gross economic inequity, in a sense fueling the desire for people to return to the local and national. The answer here lies with an amended version of "more of the same." These amendments, however, are at best greenwashing (and whitewashing) the environmental and social problems resulting from globalization and climate change.

The fourth general reaction is quite literally "out of this world." The transhumanists that want to put all their faith in modern, Western reductive and productive science and find ways to overcome our biological bodies or (in the case of Elon Musk and Jeff Bezos) flee the planet all together fall into this camp.[34] Though I do think space exploration is a needed and valuable branch of technology and science, it ought not be done in an effort to flee our own planet, but rather to learn more about our planet's place in the universe.[35] Furthermore, this elite vision is one that leaves most bodies (human and nonhuman) behind. What gets taken along to a new planetary home?

Who is the ideal human in the transhumanist vision modeled after? What gets preserved of humanity and the planet earth?

The fifth response is what Sarah McFarland Taylor (following Mary Annaïse Heglar) calls the "doomer dude" response.[36] This is a response seen on both the Right and the Left: the end of the world is nigh, and it is too late to do anything about it. On the right one might find the various apocalyptic preppers and even the billionaire bunker preppers.[37] These are the people who are preparing for a libertarian, every-human-for-themselves (or every-family-for themselves) scenario. The goal is to build a bunker, stockpile weapons, food, water, and other necessities, and learn how to become self-sufficient. For them the future is a return to a scenario that looks a lot like the Hollywood fantasy of the "Wild West." On the political left one finds the ecological misanthrope. The radical environmentalist that sees humans as the problem and thus the only solution is a serious reduction of human population. This vision calls for a return to some sort of paradisical relationship between humans and nature. Go "back" to the land, "back" to nature, as if we have somehow left it. Both groups mentioned tend to be demographically white and male.[38] In addition, one needs to be of a certain economic status to be able to afford the self-sufficiency called for by these doomers. I also find it more than a bit disturbing that as more women, people of color, queer folks, Indigenous folks, and other marginalized groups begin to share more political power, a white male response is, Sorry, it's too late, there is no world to save.

I'd argue that nothing could be more important than articulating further options, which focus on planetary imaginaries and new habits for becoming toward some other planetary future. If not, then we will continue to recapitulate the past by projecting it onto the future. We will have more of the same "isms," and the same bodies will be abjected in future becomings. There must, then, be critical articulations that consider the ways in which some bodies and processes are privileged over others, while some are considered abject. We can't just rearrange the furniture of our lives around the prefabricated banner of diversity and inclusion; rather, in order to meet the various needs of a changing planet in the coming years we need to remodel and in some cases demolish our world structures and rebuild multiple worlds together. I see language and time as two such basic building blocks that need to be reconstructed for our changing planetary contexts. In other words, we must reattune both concepts to our evolving planetary contexts.

Language, History, Evolution, and Hermeneutics

> *Linguistic forms are no sin—as long as they braid in their own unsaying. That is what as living events of language—allegory, paradox, symbol, metaphor in their epochs—do after all, that is the poiesis of language itself, its "making," its becoming: words stretch, crack, will not stay still.*
>
> —Catherine Keller, *Cloud of the Impossible: Negative Theology and Planetary Entanglement*

Languages help make the worlds in which we live: moving us toward certain ways of categorizing organisms and toward certain habits for becoming in the world. Language itself, as Catherine Keller notes in the above quote, is not the problem. We would not be who we are without it. The problem, rather, is mistaking our language for real things in the world as they actually exist. This has been variously called "reification" (per

Marx), "enframement" (Heidegger), "misplaced concreteness" (Whitehead), and "idolatry" or "ignorance" (many of the world's religious traditions). Haeckel himself understood this in his refusal of the existence of uninterpreted scientific facts.[39] As mentioned in earlier chapters of the book, once we know we are a part and product of evolution, there is no more standing outside of the process to glean some sort of objective perspective. This is part of what it means to be critically romantic: we realize both that humans are a part of the rest of the natural world so that human language is not the only thing that matters, and that we cannot know that world without interpretation, hermeneutics, and language. Whatever we call it, when we confuse our language with reality, we begin to create the world (and everything in it) in our own image. Both modern thought and naive romanticism are guilty of this. From these perspectives, we ignore the "abjections" (Butler) or the "remainders" (Derrida) in our concepts and language, and these things which are always and already left out of our concepts and languages come back to haunt us in the form of visible violence toward earth bodies (social injustices and ecological degradations). At the same time, we cannot just keep our language bound up within the human world; this is the problem of many postmodern lines of thought. There is only human language and thus the natural world is either collapsed into human language, or there is a "great divide" that can never be bridged between language and reality. If our language is how we make meaning in the world, and we are radically a part of this evolving planetary community, then our language and metaphors and meaning-making practices are a deep part of this planet as well. We must then find ways to continuously break open our language structures to let ignored and unheard voices enter into our circles of deliberation and concern. This is a "collective" process that happens over and over ad infinitum.[40] The work in biosemiotics, animal communication, and plant and fungi communication is helping us to realize that we emerge from a communicating planet.[41] From this perspective, we might reattune to the planetary community and understand things such as pandemics and climate change to "be" other earth voices beyond mere human voice for us to break out into new planetary ways of becoming. Along with Jack Halberstam, we might begin to understand the call to a new kind of queer, de-/postcolonial wildness of the becoming planetary. He writes,

> *Wild Things: The Disorder of Desire* explores the wild not simply as a space beyond the home but also as a challenge to an assumed order of things from, by, and on behalf of things that refuse and resist order itself. Wildness names simultaneously a chaotic force of nature, the outside of categorization, unrestrained forms of embodiment, the refusal to submit to social regulation, loss of control, the unpredictable.[42]

I am equally wary of voices calling for single, united solutions and visions as I am of those saying that solutions are not possible; and, of those voices claiming to have direct, unmediated experience of the natural world as I am of those who argue we cannot have any knowledge or experience of the rest of the natural world. These are pitfalls for any work geared toward creating a better planetary future. Just as a monotheistic vision of the world imposes too much control, so, too, does a single, scientific understanding of nature or any claims that we can't have any access at all to the rest of the natural world. Knowledge is always shifting, and there is no reason to believe that future knowledge will not be as different as our knowledge today is from that of a hundred years ago. At the same time, this does not mean "anything goes." Relativism ignores the contexts of the planet just as much as universalism.[43]

Haeckel and other monists of his era were engaged in public battles with idealism and materialism in German philosophy, and with dualisms of Christian theology. As mentioned in earlier chapters, they thought idealists and materialists both reduced the world to one side of the spectrum and thus could not account for the full reality of both sides. Furthermore, dualism posited some sort of hard split between entities (self/other, subject/object) and argued that the two realms of the material and the ideal are brought together through some sort of mechanical or spiritual connection, or are mediated through human reason. This dualism continued the battles of earlier Gnostic forms of dualism: it is either all material/chemical/biological or it is all mind, spirit, reason, or even mathematics. This no longer makes sense from an evolutionary context. In an evolutionary framework there is no "thing in itself" because life continuously evolves and we are located within that process (nothing is outside of it). What might appear to us (or to Kant) as a priori, external, or somehow essential, are really a posteriori "facts" and realities that have emerged in the process of entangled planetary evolution and that have been "naturalized" (in the Foucauldian sense of biopower).[44] Though Haeckel understood something like this, he still projected his own sociohistorical location onto the natural world, thereby reifying the world into his own image. We now know the violence such reifications have caused, so why not opt for multiplicity in our planetary imaginings? The argument is specifically for *planetary* multiplicities and not just the multiplicity of some postmodern relativisms, which end up being many relatively isolated entities organized by some predeveloped container. A CPR calls us to think and do something more. To critically and planetarily reattune to the worlds in which we live takes something more like what Walter Mignolo wrote of the ideas, beliefs, and practices of the Zapatistas.

> The Zapatistas, aware of that long genealogy of thought, mapped a decolonial cosmopolitan vision in the formula "a world in which many worlds would coexist" ("*un mundo en el que co-existan muchos mundos*"), and in doing so shifted from the universe of imperial cosmopolis to the pluriverse of decolonial cosmopolitanism. The Zapatistas' dictum is not pointing toward a "multicultural cosmopolitanism" where the G7 will govern and the rest will dance and eat ethnic food. To get there, to a world in which many worlds will coexist, all parties involved have to engage in the politics of decolonial cosmopolitanism, and, today, such a project could begin by confronting the imperial and colonial differences that divide and rank the world, among people and regions, as well as natural and cultural resources.[45]

A pluriverse or multiverse requires that we admit to a bit of agnosticism at the edges of our experiences. I think it is of utmost importance in the face of the urgency of planetary climate change and all of its problems that we make room for unknowing, for deconstruction, for the queer art of failure (Halberstam), and for play.[46] Failing at the language and scripts we are supposed to adhere to, deconstructing the connections between language and reality, and unknowing the things that we hold to be "true", "good," and "natural" are important if we are to stay attentive to the ever-changing planetary community rather than reify that community into our own idea of what the future ought to be. At the same time, we have to find ways to engage those who still seek a single monological truth because the expediency of that message is way more efficient than dealing with complexity and uncertainty. How can we cultivate a community that is open to multiple possibilities for planetary becoming, rather than unifying us all under a common goal, direction, language, or understanding of progress at a time when there is a rise in fundamentalist monological thinking? How can we transition goal and direction-oriented meaning-making practices, often tied up in linear, monological, and progressive understandings of "time," to emergent possibilities for the future of our planetary community? I obviously don't have an answer to this question, but I do think scrutinizing the modern Western obsession with chronological time is a helpful exercise.

Planetary Times: From Progressive Movement to Palaver

In rethinking the nature of "time" I think it is helpful to return to Haeckel for an illustrative point. One of the debates happening in nineteenth-century German thought was over whether evolution was progressive. For some, including Schopenhauer and Nietzsche, time was not inherently progressive and required sheer will to make the future better. For others, especially in theological circles, evolution and time were still progressive in a teleological sense. Though Haeckel and the monists rejected teleology, they still adopted some notion of progress within evolution: even though there could be regression (which is why education and eugenics were important for many monists, among others), overall the trend was toward more and more advanced cultures and societies.[47] Such a "progressive" understanding of evolutionary time became part of the reason Haeckel (and other's) ranked human cultures, peoples and races from "primitive" to "advanced." It was also part of the reason that many scholars beginning to compare various religious traditions at that time ranked them in a progressive evolutionary sense, usually moving from something like animism to (no surprise) monotheism. The problem is that progressive understandings of time often lead toward supersessionist thinking and to a ranking of organisms, including humans, based upon the perceived complexity or intelligence of the organism, or in the case of humans, the historical-social location of the person or people doing the ranking. The historical-social location of the observer is most often read as "the most advanced," and then read back into nature as the most progressive point within nature, not as one of many points within an ever-changing and diverse nature.

Though we may be more aware now of our own social and historical locations than Haeckel could have been, and thus of the need for hermeneutics, we still often speak of things such as "progressive" politics. What do we mean by this? Progressive for whom, what, when and where? Modernity and postmodernity are both caught up with this linear understanding of time as well, hence the pre- and post- of modernity. Again, as Mignolo questions,

> And why should indigenous people who defend the link between land and spirituality be seen as archaic, romantic, out of place, and out of time? Who is reproducing coloniality of knowledge and of being when such judgments are pronounced? Really, where do we find the warranty that the first is a modern or postmodern endorsement of truth without parenthesis and that the second is out of history, in the past, superseded?[48]

Isn't it possible that the current fossil-fueled, technological societies of the planet are an evolutionary dead end? Furthermore, isn't it possible that there are multiple experiences of time, and multiple planetary times, that cannot be subsumed by one linear or even spiraling progressive time? I think a CPR calls for the experience of many different times on the planet, not all of which fit together. As Michelle Bastian, following Deborah Bird Rose, suggests, we exist in "multispecies knots of time" that are not all run by the same clocks, and which definitely do not conform to our GMT.[49] Rather than unified time, perhaps we can begin to imagine a spectrum of time: the time of rivers, the time of glaciers, the time of humans, of trees, of forests, and of bees. Geological, evolutionary, and universal times are all of different scales. Time, it seems, is just as pluralistic and related to our embodied perspectives as are our language and concepts about reality. And time is not always linear. To imagine linear time, particularly in the case of progression from one point to another in time, many other actants and times must be backgrounded in order to highlight said time. Linear and progressive understandings of time do not, in other words, allow us to reattune to other evolving organisms around us, but rather these notions of time tend to cast every other living thing in one's own theater production called "progress."

Here I think it is helpful to think with Deleuze and Guattari about a more "rhizomatic" understanding of time, one that meanders and can move in multiple directions, rather than an arboreal understanding of time: that which progresses from root to branches, leaves, and fruit and flowers.[50] Or we might think with Isabel Stengers about time as a palaver: a meandering and nongoal-oriented walk, rather than moving in any certain direction.[51] Perhaps it is even helpful to think here of time as more like a starling murmuration: there are clear patterns, and the various individuals are clearly shaped by the movement of time, but there is no clear direction as the patterns shift and evolve into one another. Perhaps the goal is to trace the various patterns of time without necessarily forcing all times into one, single pattern, but rather just wonder at the series of beautiful patterns made by nature naturing. The time of progress may be too neat and focused to capture the various experiences and meanderings of time on a planetary scale.

These meandering metaphors allow for multiple experiences of time, none of which can be subsumed into the "tunnel of time" that progressive time forces upon the planet. It allows for us to be critical of our own sense of time while acknowledging the many different times within the planetary community, without thinking we can fully understand those times in some naive sort of way. Multiple times interact and intersect on a daily basis, and this is part of an overall ecology of time that might help us to pay more attention to particular organisms and places and worlds within the planetary community than any single understanding of time allows. Every scale of time is made up of other scales of time, and in turn becomes parts of larger scales of time. In other words, time cannot be reduced to the human scale or idea of progress anymore than nature can be reduced to chemicals, genes, or atoms. Planetary times do not conform to GMT.

Here at the end of this book, then, something like a CPR understanding of nature that is truly nonreductive begins to come into view: nonreductive in the reductive scientific sense, but also not reduced to the incorporation into the expansive self of some naive romanticisms, or the linguistic turn of some postmodernisms. If the reductive and productive model of science is just a small blip within modern science, then we can begin again to articulate nonreductive models of science and pay more attention to the nonreductive models that are already out there. I think we can learn much about these nonreductive models and the mistakes of such nonreductive naturalisms by thinking with Romantic predecessors such as Ernst Haeckel. It is here, finally, at the edge of our reflections that we can begin to talk about the emergence of a nonreductive CPR for the earth's future.

6

THE EMERGENCE OF A CRITICAL PLANETARY ROMANTICISM FOR THE EARTH'S UNCERTAIN FUTURE

Perhaps the hardest part of placing human beings into an evolutionary and ecological context, fully into the planetary community, is placing our ideas, desires, hopes and dreams within that structure as well. As mentioned throughout this book, once humans are placed into an evolutionary context, hermeneutics become key to understanding the world: We are no longer objective observers of the world, but participants in and perceivers of the world. Likewise, we cannot subsume the whole world into our own thought-system, as if we have direct experience of human and earth others. This goes also for our hopes, dreams, desires and political imaginations. A real political and intellectual ecology means that we can no longer pretend to manage a closed system, or that that system will somehow conform to the linguistic and ideational worlds that we create for it, but that we must remain open to an unknown and open future of which we are but a part. This requires not just an evolutionary and ecologically grounded political imagination that places us into a planetary context, but also a nonreductive naturalism, or what I am calling a critical planetary romanticism (CPR) that grounds our epistemologies and ontologies.

Some of the nineteenth-century Romantic scientists such as Ernst Haeckel understood some of these types of insights, yet most were incapable of fully placing their own ideas within an ecological and evolutionary context. Instead of understanding his ideas as options among many others, in different evolving, open-ended contexts, Haeckel ordered all of life and all human cultures into a hierarchy that led up to what he thought was the most advanced and progressed life at that time: European culture. He was, so to speak, looking down the well of geohistory to see his own image reflected in the water below. This had all sorts of racist and ethnocentric implications as well as dangerous implications about eugenics and the perfection of the human race that would come to haunt science and humanity in early twentieth century[1]

Many scientists (among others) today still believe that whatever knowledge we have now is the "best" knowledge possible and that other ways of thinking should be judged by the latest ideas coming out of the modern sciences. The political cry "Science is real" in the face of fake news and alternative facts is one example of this enduring assumption, as if other ways of thinking cannot also be real. Some environmentally minded folks are equally certain, believing that they can delineate between what is natural and unnatural and manage the world accordingly, even if that means displacing or ignoring Indigenous and local peoples' knowledge to do so. On the other side of things, some forms of postmodernism argue that we are so trapped in human language, that we can't "get to" any descriptive experiences of the rest of the natural world, or even worse, we can't get to any reality outside of some type of identity politics, language, or culture. These are reductive attitudes toward the world, based on the idea that we can reduce the world to human ideas, visions, goals, or measurements. Rather than forcing all evolving life into one's own perspective, or even some vague human form as the era of the "Anthropocene" might suggest,[2] we might be better served by placing all human thought and ways of thinking within an open-ended evolving, ecological context. We might call this, along with Donna Haraway, the Chthulucene in order to highlight that emerging and thinking is always emerging and thinking with others as hybrids, or that autopoiesis is always sympoiesis.[3] But whatever we call such an evolutionary and ecological understanding of thoughts, values, emotions, desires, hopes, and dreams, I think we can draw from the ideas discussed in previous chapters about the queerness of nature and suggest that from the perspective of a CPR mapping out an "ecology of thought" should embrace uncertainty, ambiguity, plurality, and failure. Uncertainty and ambiguity simply imply that all of our knowledge is insufficient and that it opens on to an unknown world that at some point cannot be contained by it. This is the result of being a part of the planetary process rather than in a position above or outside of it. Technologies and policies of certainty do not allow for openness and ambiguity and thus cut living entities off from evolving planetary communities. Instrumental technologies that create certain results do so to the neglect of the complex ripple effects a given technology or action has on other entities. What might technologies and policies of ambiguity look like? What might technologies for the planetary community look like?

Plurality simply recognizes that our own perspective, or our own cultural zeitgeist, is always one among many, and is located geographically, biologically, culturally, and historically. As Catherine Keller, Laurel Schneider, and others have argued, polydoxy not orthodoxy is the norm.[4] Technologies and policies based on abstract objectivity and universality cut us off from the multiplicity of evolving contexts. Most of us moderns, for instance, don't even know where our daily food and water comes from: we are cut off from our local contexts through the daily technologies we use.

Assumptions based upon ambiguity, uncertainty, and plurality require us to also embrace failure.[5] Sometimes when we exist and persist in a system that is destructive or detrimental to a great deal of life, then the best thing we can do is fail that system. We need "false thinking" and "wrong headedness" to get us out of the trap of reducing our ways of thinking to business as usual. Becoming experimental in our thought might help us to, first, see how our ideas materialize in better and worse ways as they ripple out through the ecosystems and planetary communities, and then, second, adjust our thinking and actions accordingly.[6] Technologies of progress and success lead to the backgrounding of the unforeseen negative consequences of those technologies: Think about the Green Revolution and the unforeseen problems of industrialized agriculture, or renewable energy and the unforeseen problems of mining lithium and other resources needed for batteries. What might technologies and policies that embrace "the art of failure" look like?

In an ecology of thought, ideas are not unidirectional but return to the thinker(s) with feedback loops from the larger ecosystem(s). From within an ecology of thought, religion might be best understood as a constant process of critical reattunements to the worlds around us, with attention to how different bodies are affected by different co-constructions of the worlds in which we find ourselves. To explain further what a CPR might look like, in this final chapter, I first highlight some of the failures of reductive and productive models of materialism.[7] In doing do, I look first at Haeckel's own hubris and lack of humility. Hubris and lack of humility are sure signs that one is reducing the world to some common denominator. Then, I examine the problems of certainty in both religious discourse and in scientific discourse. These two "anchors"—the scientific and religious—help to fuel the lack of humility that leads to ending dialogue. They create what is sometimes in epistemology called foundational concepts or "conversation stoppers." As we confront our own hubris, and the ways that they are anchored in scientific and religious discourses, then we might begin to open to an outlook that promotes multiculturalism and multinaturalism. One cannot exist without the other, as Bruno Latour has, I think, persuasively argued.[8] As such, we might need to shift our desires for justification from the past and the present (looking to historical-textual and evolutionary-biological records and data) toward a justification from the future: a justification that is in a Derridean sense always and forever "to come."[9] Just as we are thrown into an evolutionary and historical context that is not of our own making, so the future is not of our own making. Our lives, and the planetary community are forever open to a changing future that recontextualizes and reinterprets the past: for better and worse. This insight is too much for most of us to live with, and so we project all sorts of certainties onto an unknown future. However, such projections fuel certainties that reduce the world into our own hopes, desires, dreams and imaginations. And it turns out this forcing comes at a great cost to many planetary bodies. Ernst Haeckel was just as guilty of these projections as anyone else, and it is to a discussion of his lack of humility (as a warning to us all) that I now turn.

WHAT WE CAN LEARN FROM RELIGIOUS CERTAINTY

Catherine Keller argues, in her book *God and Power*, that certainty has perhaps caused more problems and violence than uncertainty ever has.[10] Agreeing with this sentiment, I would argue that the majority of religious violence stems from the closure (epistemological, metaphysical, ethical, aesthetic) that certainty enforces upon the planetary community. It is not hard to see examples in the daily news: religiously fueled racism, sexism, ethnocentrism, heterosexism, religious bigotry, and general hatred toward other peoples flood the pages and live feeds of any given news or social media source. Deep disagreements between the so-called Left and Right also come from, I would argue, deep certainties on both sides that are beyond what the human mind and human reason could possibly ever substantiate. The ways we deal with and react to a global pandemic and to the effects of climate change have shed light on just how uncertain and precarious the future is. The certainty on both sides of recent controversies (vaccinate/don't vaccinate or wear a mask/don't wear a mask), turned into political litmus tests for whether you were on the left or right in the United States. Oddly, the certainty on the far right in terms of natural rights and individual freedom and the certainty on the far left in terms of what is "natural" meet to oppose vaccinations. Certainties lead to these types of strange political bedfellows in a politics of piety.[11] Even with such odd agreements, this type of certainty doesn't allow for constructive dialogue that leads anywhere. There is a similar example around issues of climate change and its effects. I'm not sure whether politically denying human-caused climate change (on the right) and shouting "Science is real" (on the left) are anything more than shouting matches from places of certainty. Science is, of course "real," and so are values, ideas, beliefs, affects, and all the other things that make up human beings. The Enlightenment assumption that we will all be compelled to adopt a single and unified true knowledge and change our behaviors accordingly doesn't seem to hold true. Certainty breaks apart on the shores of our evolving planetary realities.

If we consider that certainty is the opposite of hope and faith (for certainty requires neither of these), then these epistemological, philosophical, ontological, or theological certainties signal a hardening of ideas and values toward instrumental, individual ends that suit one's own needs, ideas and desires. How can one be open to constant reattunement to the changing worlds in which we live with a mind and attitude that are closed off to newness and otherness under the banner of certainty? This is the opposite of what religious attitudes are intended to do (to constantly reread and gather together all existing information in a meaningful way), and it represents what we might call a modernization of the religious mind. In other words, many within religious traditions (lay, practitioner, cloistered, and clerical) have taken on the modern understanding of a singularity of certain truths that can be known by all humans. Truths, in religious traditions, are always couched in some sort of mystery that humans can never fully understand. Even in monotheisms, one finds apophatic, mystical, and iconoclastic thinking. This is one of the core themes in most religions. Namely, I would argue, religions are primarily about opening us on to "others": beyond self, beyond kin, beyond nations, beyond species, and eventually toward the planetary community. There will likely be a future where we will challenge ourselves beyond the earthly community, but until such contact is made with planetary beings outside of Earth, it seems that the scale of the planetary provides a good context for our ethical and aesthetic imaginations. Indeed, hundreds or thousands of years from now, a "planetary" understanding may be seen as out of touch as the idea that the universe is six thousand year old to many of us today; but for now this earthly context is, I argue, the scale we desperately need to think with in this moment.

To be open to "the other," it helps to admit a certain amount of not-knowing and uncertainty. This may be one of the primary reasons that all religions are geared toward producing experiences that lead us to wonder and marvel about the worlds around us. As Mary-Jane Rubenstein writes,

> Insofar as wonder can function as a kind of wound in the everyday . . . it must again be emphasized: just as a wound ceases to be itself when it heals, wonder is only wonder when it remains open. Wonder opens an originary rift in thought, an unsuturable gash that both constitutes and deconstitutes thinking as such.[12]

This "keeping open," the space of mystery, wonder, the *tehom* of Geneisis, the *chora* of the Greeks, the *neti neti* of Hinduism, the *anekāntavāda* of Jainism, the apophatic and negative traditions found in Sufism and other forms of monotheistic mysticisms, the trickster figures of many Indigenous traditions, and death of God and Western deconstructive philosophies are all precisely what enable the emergence of new ways of becoming in the world. The great religious stories of Moses, Jesus, Mohammed, of the Buddha, Lord Mahavir, stories from the Bhagavad Gita, and from Confucius's writings are all largely geared toward producing wonder and openness toward the miraculous in everyday life. The folly of totalization and certainty are often targets of these stories. Finally, many rituals such as meditation, yoga, contemplative prayer, chanting and singing, dancing and drumming, and shamanic journeys are largely geared toward getting us out of the rational mind and opening us up to wonder and mystery. The open spaces created by these ideas, stories, and rituals make necessary our constant reattunement's to the worlds in which we live. To attune in ways that do not involve naive projections of certainty, objectivity, and universality, we might do well to remember the unknowing at the edges of all of our knowing.

Part of the hardening of knowledge systems into metaphysics has to do with the desire to attach too much certainty to our interpretations of the world. From an epistemological sense, one can argue that whatever else "modernity" is, it is this fixing of ideas and reality into a 1:1 relationship, whether these ideas be scientific, religious, philosophical or other. This happens especially when knowledge claims are affixed to the natural world or what one thinks is "natural." Once ideas are read into the natural world, powerful inertia builds around these ideas and begins to create the world in these power-filled images of what the truth is. This is just as true of the Christian church as it is of terrorists who kill in the name of what is certainly (for them) right and good, as it is for those who argue that modern, Western science will solve all the riddles of the world and give us the one true understanding of the world.

In the case of the Christian church's slide into modern certainty, wonder, as Lorraine Daston and Katharine Park have pointed out, became a source of fear for the church, a challenge to the received authority of hierarchically produced dogma.[13] Wonder was the beginning of curiosity that would lead one too far down the path of the senses and individual hubris. Theology, in the Augustinian form of fourth-century empire theology, for instance, read the Trinity into the Ptolemaic/Aristotelian universe, all brought into existence ex nihilo by the unmoved mover creator God. Christian Monotheism, despite the Trinitarian acrobatics necessary to describe how a transcendent omni-god could appear in a changing, embodied, material form, succeeded for a time in fixing creation itself into the idea of a single history ruled over by church hierarchy: the technology of the church proved to be pretty powerful in the process of colonizing Europe and other continents.[14]

This fixity was, of course, never complete; there were always voices counter to this dominant "empire" reading of religion. Again, polydoxy is the norm. Many of the early so-called church fathers even found mystery in the concept of *creatio ex nihilo*, which went against the ancient Greek idea that nothing from nothing is nothing.[15] The mystics of many traditions, and within Christianity, St. Francis of Assisi, Teresa of Ávila, Nicholas of Cusa, Giordano Bruno, Bartolomé de las Casas, later liberation theologians, and many more all challenged ossified and naturalized readings of Christianity.[16] These thinkers challenged the fixity and certainty of theological dogma while natural philosophers, such as Thomas Aquinas, were beginning to argue for the greater certainty of knowledge one gains through direct study of nature. The challenge of metaphysics on the part of some theologians was based upon the art of unknowing or wonder that allowed them to challenge received tradition and think anew about the world around them. Catherine Keller notes, for instance, that "what unfolds in the cloud-space of the Cusan God is a multiverse of perspectives, proliferating, holographically, irresolvable into any fixed proposition."[17] Again, whatever else a living, changing, universe might possibly be, it is precisely something that is forever in relation: changing as it folds in and out of relations with other enfolding and changing entities in a seemingly infinite process. Certainty claims too much when it becomes absolute certainty, in essence denying the evolving entanglement between and among entities. As Louise Economides puts it, "To wonder is to wander. To not only tolerate uncertainty in the Keatsian sense of 'negative capability,' but also welcome it as a gateway to new possibilities. To not stay put."[18] When we deny our shifting entanglement in support of fixity and isolation, we also denying the sensory embodied exploration of the world: that constant reattunement to our worlds that is required of us as living entities. This is, again, one reason that Haeckel and other Romantics argued against a priori knowledge and argued for an educational reform in which every student spent more time within the natural world, exploring the world for themself.

Copernicus and later Galileo (following Bruno and Cusa), like Haeckel and the Romantics a bit later, began to explore different possible explanations of the cosmos, in part because they began to trust again in their embodied, sensory explorations of the world, enhanced by new technologies that led to questioning received dogmas. Such questioning eventually led to a transformed cosmos that then caused the likes of Descartes to question all foundations and even all sensory information. This deeply deconstructive, negative form of knowing, more akin to an epistemology based upon the Buddhist concept of Nothingness, Emptiness, and Form, meant that received understandings of God that fixed the foundations of the universe could no longer be assumed.[19] The Protestant Reformations further challenged the theological authority of the church. However, unknowing and deconstruction proved to cause too much anxiety. Descartes would posit the foundational cogito to prevent the slippage of knowledge into an infinite regress, just as the emerging scientific method would transform the Revelation of God shaping all things created and known through reason and faith into the "natural laws" achieved through the scientific method and reason.

In many ways the story of Copernicus and Galileo would become the beginning of a progressive story in which the natural sciences provided the only method that could give us better and better understandings of reality, and Haeckel and other evolutionary thinkers (and scientists in general) would pick up on this progressive thread. So, rather than a fundamentally deconstructive moment in which new ways of knowing and understanding the world could begin co-constructing the world in a different way, the religious elements of metaphysical certainty were adopted from a specific reading of monotheistic Christianity (and Islam) into the emerging scientific worldview. The Christianization of the world, in other words, simply gets replaced by the Enlightenment narrative, in which the light of certainty in science (guided by reason) brings us out of the darkness of religious and parochial dogmas, into the one truth of the natural world. In addition to critiquing religious certainty then, it's also important to critique certainty within scientific discourse.

WHAT WE CAN LEARN FROM SCIENTIFIC CERTAINTY

As the certainty of theological and religious understandings of the world was challenged through the mixing of ideas about reality made possible by travel, trade and colonization (now globalization), and by advances in so-called scientific understandings of nature, many scientists and those who followed the emerging scientific worldview assumed that the uncertainties of religious and philosophical ideas about the world would be replaced with the certainties of a scientific method. The mono- of monotheism was imported into science as universal truths about the natural world discoverable through experimentation, investigation, and reason. Something within human psychology yearns for the type of certainty that this (modern) monological thinking (falsely) promises. How can we act in a world that is uncertain and in which we cannot understand the full consequences of our actions? The thought of infinite regress is just too much to bear at times. However, uncertainty and regress, to some extent, seem to be written into our experiences of the world and cosmos of which we and all our thinking are but a part. We cannot see or experience the whole, and hence the reason that any Romanticism must be critical.

Science experiences radical shifts in understanding (even paradigm shifts) from time to time in such a way that the scientific story of a hundred years ago, of Haeckel and his colleagues, for instance, is not the same scientific story as today. Most scientists know this and wouldn't argue with the premise. We now have knowledge about genes, that the universe is about 13.7 billion years old and expanding in all directions, and about subatomic particles that Haeckel and the scientists of his day did not have access to as they were building a "naturalistic" worldview. All of this and everything else we know about the world is based largely upon the 5 percent of the visible universe; the other 95 percent of the universe made up of dark energy and dark matter we know almost nothing about. Making universal statements based upon 5 percent of the visible world, from a tiny dot we call planet Earth, and in a universe that has been expanding in all directions for 13.7 billion years, is bold to say the least and might better be described as hubris and folly.

Furthermore, we often argue that the present state of knowledge is the "best" and "most accurate" of any given era (at least from the modern Western perspective). This presentism supports a belief in progressivism. Now is always better than before, and that will continue ad infinitum. However, as cosmologists remind us, if the universe is expanding at a faster and faster rate in all directions, then at some point, the galaxies and stars we see at night will be so far away that we will only be able to see

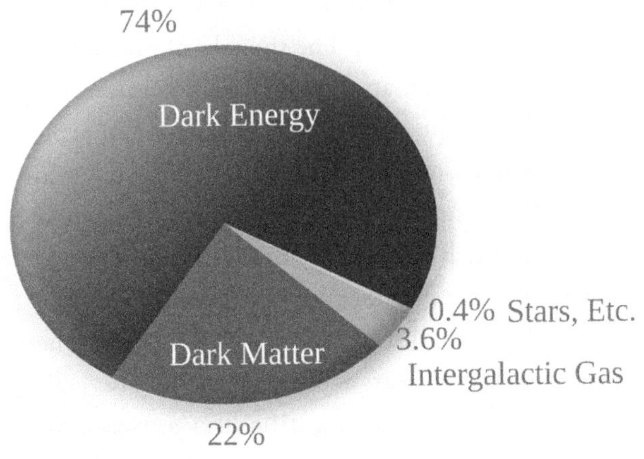

FIGURE 6.1 Dark matter, dark energy, and the visible universe. (Courtesy of Wikimedia Commons)

the light from our own galaxy.[20] Those looking out into space in the far future may think our own Milky Way galaxy is the only one that exists and hence perhaps a return to the idea that humans are the center of the universe. Our knowledge is context driven, and to mistake the present as inherently progressed from the past is a mistake. We are thrown into contexts not of our own choosing to co-create the world though specific technologies that then co-create the stories of the worlds in which we live. This co-creation and constant change is not a demotion of the project of knowledge production: it simply means that there can be multiple valid perspectives from different embodiments on planet earth at any given time, and that these knowledge claims are always only valid for a time. The nature of the type of knowledge we can co-create is not universal, unchanging, progressive, and lockstep; rather, it is contextual and in flux as we reattune to the changing worlds around us.

And still, we humans can't help but to try and understand the context in which we find ourselves and develop criteria to navigate between "better" and "worse" understandings. How best do we do this without the crutches of religious and scientific certainties? I find the start of an answer in Catherine Keller's *Cloud of the Impossible.* She argues, "If something is known at all, it cannot be absolved of relation; therefore, nothing is known absolutely."[21] In other words, we must become comfortable with knowledge that has built in unknowing. This is what the "re-" of *reattuning* and *religion* is all about; holding open that space, even if parenthetically, so that we can form a ground solid enough from which to act and make decisions.

Every scientific shift in our understanding of the world depends upon some form of unlearning, failure, and apophasis. Cosmological and scientific shifts provide fruitful examples of the ways in which not knowing is always intimately tied to that which we know. The Copernican shift in the fifteenth to seventeenth centuries, the evolutionary shift in the eighteenth to nineteenth centuries, and the ecological shift in the twentieth to twenty-first centuries exemplify this point. These co-constructions or shifts help make the possibility of planetary communions central to *commun*ication and the possibility for making planetary conversions central to *convers*ation, precisely because their relationship to the world keeps knowledge claims open to the shifting terrains of the various evolving embodiments that make up our planet. If, after all, all of knowledge is a foregone conclusion, communication and conversation are not possible: the only option is acceptance and assent.

If one follows the narrative of a progressive understanding of knowledge and in particular scientific knowledge—as in science and the sciences are bringing us closer and closer to truth—one must assume that reality is stable and that our reason and capacity for understanding the world is based upon a likeness to the laws of the universe and of evolution. This model of progress, argues Karen Barad, follows a Heisenberg-influenced understanding of truth in which our scientific models' inch closer and closer to the truth as we get better and better knowledge. In other words, communication is not really any type of communion or conversation (both of which imply the ability to be changed by a dialogue) but rather acceptance. What *would* happen, Barad wonders, if we were to follow the epistemological implications of Niels Bohr's assumption about an indeterminate reality?[22] In this model, rather than knowledge continuing in a long line of colonizing progress—Christianizing or "enlightening" all that it comes into contact with—knowledge is a co-constructive process that creates different truth regimes or habituses for becoming in the different worlds in which we live. From this perspective, the question turns from how true our ways of becoming are to how different embodiments are affected within a given truth regime and different habits for co-constructing the worlds in which we live. In other words, our concern with knowledge moves away from metaphysics and ontology and toward aesthetics and ethics. This, I argue, depends upon the space of unknowing and undoing that is found in many religious and philosophical traditions, which is also built into the scientific method itself as it exchanges one theory or model for another that better accounts for more data. It depends upon a certain movement of knowledge and shifting of certainty that allows for us to be surprised by what the other (whether human or nonhuman) has to communicate.

In the end, from this understanding of knowledge as always uncertain, we might accept some form of what Fuyuki Kurasawa calls "cosmopolitanism from below," and some form of what Bron Taylor calls "terrapolitanism."[23] By "from below" I mean thinking with and from the margins, from borderlands, from the "third spaces." In addition, terrapolitanism adds that these spaces are not only human, but involve the entire planetary community. Bruno Latour's idea of thinking with and within a "critical zone," which is basically the biosphere, and Vandana Shiva's political vision of an "earth democracy," which includes the nonhuman world in a planetary form of democratic politics along with the human, are also helpful here.[24] Such thinking helps to guard against projections of certainty in the world that would, for instance, rank all life in a hierarchy of value, or suggest too close a connection between progress in evolution and a particular, culturally located scientific understanding of progress. This is the type of thought that thinking from the "top" or the "center" creates. The complexities of a planetary community do not allow such simplistic understandings of our socio-natural worlds, and in fact, expose them as deeply rooted in mono-theistically influenced cultures.

POSTMODERNITY AND (COSMO-) TERRAPOLITANISM "FROM BELOW"

The "postmodern" shifts that took place in Western Europe were in large part a correction of the certainty imposed by modernity. As such, people began to fall from high places of objectivity to the grounds of the earth. With that (re)grounding, people began to kill off the false idols of transcendence: God, the Self, the Author, the Text, binary gender, and so on. This regrounding or making immanent of all things human (including thoughts and imagination) also led to a proliferation of perspectives: There is not just one story under which we all fit, but rather multiple stories, and with them multiple truths.[25]

These shifts were also the result of activist movements that challenged the oppressions caused by interpreting history from a single tradition or single perspective. Women's suffrage, civil rights, the farm workers movements, Indigenous rights movements, queer movements, and liberation struggles of various kinds against old colonial empires all led to the questioning of a single reading of history and called for an end to the oppressive and violent technologies that emerge from such a reading. Each one of these movements challenged dominant received knowledge and interpretation of a single truth, history, or worldview. In addition, the horrors of World War II laid to rest any facile notion that the modern, Western industrial world was on a progressive course.

To examine our present, à la Foucault, from postmodern grounds we must go back into the detritus of history and examine the different layers of sediment to see how our understandings of the world were co-constructed over time. The postmoderns have been able to examine the various politics of truth-making, and this has led to a proliferation of critical theories, many Marxist-based, and many based in particular identity politics to address specific societal oppressions. Feminist, Black, disability, class, Latinx, queer (though critical of "identity politics"), postcolonial, and other critical theories began to challenge the idea of a normative human, of the idea of a natural "race," of the idea of stable genders and sexualities. Each of these critiques widened the tent of possibilities for the becoming of bodies and of human views (*theoria*) on the world. However, many of these critical theories were (and are) concerned mostly (if not solely) with the human world, with some form of human pluralism that involves the widening of human perspectives and possibilities. Many were (and are) also quite wary of the concept of "nature," if not against it all together, because nature, like god, has been (and is) used as an anchor for various oppressions, that started with some form of the sentence: X is bad because it is unnatural (or ungodly, or both).

By the late 1960s, after two world wars and an ongoing Cold War, with the publication of Rachel Carson's *Silent Spring*, the famous "Earth Rise" image, and the growing awareness of environmental issues around the world, it was clear that the modern mindset—and its (never completed) death of nature by the reductive and productive model of science—was a huge problem. It also became clear a little later that postmodernity, especially of the more constructivist variety, wasn't offering much in the way of an environmental ethic. There are, of course, exceptions, and these are now growing, such as David Wood's idea of "e-construction" and the constructive postmodernisms such as found in process philosophy.[26] Even more recently there have been a host of "immanent" ways of thinking that are returning human thought to the *humus*. It is with some of these that I think we are beginning to see a "third" option that breaks us out of the (Western) human chronology of time moving from pre- to postmodern/colonial and offers instead a critical planetary romanticism based in embodied eros, or erosophia.[27] I'm aware that I am also telling a narrative here, but it is *a* narrative, not *the* narrative and it doesn't rely on a single tunnel of progressive time. A million other narratives and stories could blossom in distinction from this one, and all of them together make up the messy intertextual/-sectional planet in which we live.

It is commonplace to talk about pluralism in regard to understandings about human cultures, ideas and values, but not so common to talk about it in terms of our views of nature and reality. As Latour (among others) has pointed out, for the moderns, nature appears as the objective space in which ideas about it are anchored.[28] And, I would add, for some postmoderns it appears as a blank space inscribed by language. However, placing humans within an evolutionary perspective and all that that entails also means that we must begin to think of multiple natures if we are going to admit multiplicity among human beings. This does not mean relativity, but rather it is simply suggesting that "nature" is the relationships within and between organisms that make up a given place and time. The entire earth (and all entities that make it up) is agential, not preformed and static, nor passive and awaiting inscription. As Latour reminds us, "Two of Gaia's surprising features are, first, that it is composed of agents that are neither deanimated nor overanimated and that, secondly, contrary to what is often claimed in criticism of Lovelock, it is made of agents that are not prematurely in a single acting whole."[29] If this pan-agential, relational reality is the case, then not only can we talk and think about multiple *ideas* of nature, but also multiple natures: shifts that occur between organisms and how they relate to one another throughout time and space. These are not just shifts in perspective, but they change the ways we relate to the rest of the natural world and thus change the course of nature itself, just as different natures change what it means to be humans in relation. Think about how a reductive cosmology helps to shift natures toward the technologies of the Industrial Revolution, or how eugenics and genetics have co-created technologies that have led to shifts in natures. Consider how pandemics, viruses, and the microbiome have codeveloped the vast variety of beings that fall under the banner of human, and how the various plants, animals, geologies, bacteria, viruses and other earth entities have changed the course of humans and the planet. The point here is that the recognition of pluralism in nature and culture go hand in hand and are coproductive. So we need some sort of understanding of multiculturalism and multinaturalism, neither of which can be founded in transcendent principles.[30] From a CPR perspective, we can recognize both sides of the nature-culture Mobius strip and recognize that both have co-constructive histories that can be excavated and examined. What we have, as Haraway has pointed out are "naturecultures," or what the theologian Gordan Kaufman called "bio-histories," or what Latour calls "geo-stories."[31]

Though many reading this book may take ecological entanglement and relationality for granted, such entanglement and relationality was hard fought for conceptually from a modern Western context, and is still being fought for in terms of how it plays out in our materialized bodies. Relationality and entanglement were literally beaten out of many people during the period of European colonialism. The imposition of individualism, nuclear Victorian families or plantation hierarchies in the case of slavery, heteronormativity, and gender and sex dimorphism were all part of the "technologies" that worked to conform human life to the confines of the euro-Western understanding of the world. Most Western moderns, I would wager, still probably don't think of their own "self" as a microbiome, as an ecosystem, even though there are ten times as many "nonhuman" cells in our body as there are human cells. At the same time, more and more are refusing the reductive productive forms of science and opening up to the ways in which we humans and our technologies are merely a part of the rest of the natural world. I think there would be no Donna Haraway and her linking of technology-humans-and-nature, no Bruno Latour's collective or science and technology studies, no understanding of bio-history, without the evolutionary work of people like Darwin and Haeckel; and there would be no robust theory of evolution without some sort of Romantic movement influenced by the influx of Vedic, Indigenous, Islamic, and other relational understandings of reality in Europe during the eighteenth and nineteenth centuries (and a bit earlier in the case of Islam), which of course depended upon the travel and interactions brought about by the European era of colonization (and earlier trade routes between China, India and what is now the Indonesian archipelago). Some of the strands of planetary thinking also include events such as "the great acceleration" and the "Earth Rise" image. The point here is not to suggest that the present "is the way things are," nor to suggest that the ways we understand ourselves is inherently any "better" now than it was or will be. Rather, I am only trying to give some context and benefits of a CPR way of thinking.

CRITICAL PLANETARY ROMANTICISM

Other species, as well as nonliving things, make it possible to be human. That this statement is not obvious to many humans at this moment in time is only because of habits of thought that have become powerful over the last few hundred years. In this modernist mode of thinking, humans presume to transcend and master nature, rather than forming worlds together with nonhumans. One significance of recent discussion of the Anthropocene as a time of human-sponsored environmental crisis is that it urges us away from those powerful habits of thought. The imagined mastery of humans no longer looks so successful; we are asked to reconsider the ways in which human and nonhuman histories are inextricably intertwined. The Anthropocene, like every other trajectory in which humans have been involved, is more-than-human.

—Anna Tsing, *Feral Atlas*

There will never be a single framework to "rule them all." That desire is just a repetition of the modern Western tendency toward universalization. As Gayatri Spivak, who first used the term *planetarity* in a similar way to the way that I use it, noted: The planetary is opposed to the global view from above.[32] The planetary is simply a collection of the plurality of worlds that make up the earth at any one moment in time, both human and non. The planetary at any given moment is, as Anna Tsing suggests in the quote above, what makes being human possible. It should not be thought of as an overarching container into which worlds fit, but rather its shape takes on all the evolving relationships and bodies that make it up at any given moment. With a slight modification to Walter Mignolo's decolonial ideas, there are many worlds, at any given time, which can link together to make up the planetary. Practicing a CPR means listening to as many perspectives as possible within the planetary community: from this perspective the more perspectives we listen to and account for, the better understanding we have of the planetary at any given time. Here I want to describe further each of the terms of CPR, and I begin with romanticism.

Romanticism. As some have argued, the Romantic movement in Europe during the eighteenth and nineteenth centuries was in response to the fallout of industrialization and the beginnings of the fossil-fueled era. Some also argue that European Romanticism contains the vestiges of what was left over in the Indigenous local traditions stamped out by the spread of Christianity.[33] Indeed, Romanticism then and now does have many similarities with older Indigenous traditions worldwide. According to most animistic traditions, humans are interrelated with other animals and the earth, which is often considered sacred. Various Indigenous peoples interpolate their culture, language, and cosmology with features of the local landscape. In Jogjakarta, Indonesia, for instance, the Javanese cosmology is tied up with the volcano Merapi (the male god) on one side of the region and the Indian Ocean (the female goddess) on the other end of the region. Located right in the middle is the Kraton, or Sultan's palace, and the reigning Sultan is considered to be married to the goddess. There are several sacred places lined up between Merapi and the ocean, some of which are marked by obelisks, whereas others are sacred water spots or trees. To learn about Javanese philosophy and cosmology, then, one must walk through the geography of the place. This is one vision of what it looks like to reattune our bodies to the earth.

There are Romantic strands in most mystical traditions of monotheism as well. And plenty of earth-affirming/worshipping traditions in just about every religion we might categorize as a "world religion." Perhaps it is high time we highlight these nondominant strands as we reinterpret ourselves as part of planet earth. Science, too, has its Romantic traditions and proponents. Goethe's *Metamorphosis of Plants* and Erasmus Darwin's *Botanic Garden: Containing the Loves of Plants*, are both examples of this early Romantic tradition in the emerging sciences.[34] Ernst Haeckel understood humans as part of evolutionary theory and thought the study of nature produced wonder and awe. This particularly comes through in his artwork, and in one collection titled *Kunstformer der Natur* (Art Forms in Nature).[35] Wilhelm Bölsche wrote a book titled *Das Liebesleben In Der Natur* (Love Life in Nature), as did Gustav Fechner on the "souls of plants" *Nanna, Oder über das Seelenleben Der Pflanzen* (Nanna, or On the Inner Life of Plants).[36] These eighteenth- and nineteenth-century scientists argued that some form of a nonreductive understanding of nature should be the basis for the sciences. Of course, there are plenty of scientists and other scholars today that take Romantic readings of nature as scientific.[37] New materialisms, emergence theories, animisms, and other philosophies and theories have all more recently called us to a think through a form of Romanticism that gives agency and value to the more-than-human planet. Perhaps it is time we reattune our thinking to these ways of thinking. Some form of Romanticism could help re-embed the fossil-fueled humans (and all things human) into the mineral, planet, and animal worlds. However, this Romanticism cannot be local, parochial, or national, and hence the need for planetary thinking.

Planetary. One critical difference between earlier Romanticisms and perhaps animisms as well, is that there is no cutting localities off from the planetary flows of materials, energy, and information. We live in a globalized world, for better and worse, and I don't see a "return" to localism, parochialism, or nationalism as good options. Making Romanticism local provides too much of an opportunity for "us" and "them" thinking. Keeping out "illegal" immigrants and "nonnative, invasive" species results in much violence to earth bodies and doesn't reflect the worlds in which we live.[38] I'm not necessarily arguing for complete open borders and allowing all nonnative species in, as this would create its own violence. Our bodies and all the diverse bodies of the planet depend upon constraints and boundaries. But these constraints and boundaries are flexible and shifting over time. I'm arguing for a reattunement to our planetary contexts, which even before fossil-fueled globalization experienced exchanges between ecotones, oceans, and the atmosphere. Even when one stays "local," one's life is crisscrossed daily by global flows of energy, materials, and information and these exchanges eventually lead to shifting boundaries and new entanglements. Thus, I have called elsewhere for a polyamory of place: the love of many places that we can link and connect through planetary flows.[39]

The *planetary* is also a term meant to remind humans of our creatureliness. We are first and foremost creatures among creatures, sharing the "common grounds" of earth, air, fire, and water.[40] Humans are nothing without the planet. This may seem obvious, but the hopes and dreams of leaving this planet on the part of some people such as Elon Musk and Jeff Bezos means that some versions of human anthropology still suggest we are somehow not a part of Earth.[41] A focus on the planetary also means realizing that whatever we think of as "the best" of humans is a result of our entanglement and relationality with the rest of the natural world rather than in spite of it. Human music emerges from the music found in many other species, human art from its earliest phases is inspired by forms in nature, and human technologies don't emerge without earlier technologies found in other animals. What makes us human is enabled by our planetary kin. Finally, the focus on the planet cannot be singular because we are multiply embodied and experience the world differently. Therefore, we need the critical part of CPR.

Critical. CPR for the earth must be critical. There is no single experience of the planet and there are multiple worlds at any given time of the planetary processes that make up the planet. Different embodiments in different worlds experience the planet in multiple ways. We need critical theories of race, gender, queer studies, disability studies, class, post- and decolonial studies, animal studies, and affect theory in order to map out the ways in which certain worlds privilege certain bodies at the expense of others. From a CPR, these "multiple maps" provide us with better understandings of any given current construction of the planetary.[42] In this case, the more critical perspectives we can listen to, the more "objective" understanding we have of the planet at any given moment. Pluralism, multiperspectivalism, and diversity do not need ranking from this perspective; they are each different and valid perspectives at any given planetary moment.

Of course, we must also contend with "fake news" and "alternative facts" in the ever fluid and shifting world of the internet and twenty-four-hour news and information cycles. Critical perspectives do not mean "anything goes." In fact, the more critical our perspectives, the better. Uncertainty and multiperspectivalism are better judges of knowledge than certainty and monological thinking. One might think of a scatter plot of perspectives: There will be many that make up a core plotted along a graph, with a few outliers, and some completely off the chart. In co-constructing worlds and knowledge, and the actions based upon them, from a critical perspective we should listen to them all, but the extreme outliers might be especially scrutinized. I would argue that conspiracy theories, fake news, and alternative facts are based on "certain" knowledge to the exclusion of all others. The inability to change one's mind or consider alternatives is a sure sign of one's self-certainty, and in a CPR such monological, dogmatic thinking is dangerous. At the same time, those majority ideas at the center of the scatter chart should also be scrutinized: continuation of "business as usual" will also do nothing to help us solve contemporary planetary problems and may just perpetuate those problems. These centrist ideas also rely on a kind of certainty of "the way things are." Contextual, critical thinking enables us to consider "other's" perspectives (human and nonhuman), to reattune to the needs of neglected and abjected bodies, and to come to some sort of collective decisions that help us move in directions that will bring about more flourishing for the entire planetary community.

At any given time, we live in multiple co-constructed worlds that, when linked together, make up the planetary. Rather than there being a single narrative or line of history and progress, I would offer that CPR suggests the planetary is more like a starling murmuration. There are beautiful orders and patterns to be found, but ultimately reality is more rhizomatic and meandering: it flows together from one place to the next, making up a given manifestation of the planetary community. From this perspective we can create "cartographies of violence," taking note of how a given construction of a world creates violence for some assemblages and privileges others.[43] Given this data, we can work to co-construct new worlds that take into account more bodies and more perspectives in a way that promotes greater planetary flourishing. Here at the end of this book, I focus on one example of a planetary thinker found in the work of Rachel Carson.

CARSON'S WATER ETHIC AND CRITICAL PLANETARY THINKING FOR PLANETARY POLITICS

It was Carson's relationship with the sea, that watery depth of relationality and ambiguity that Keller writes of unrelatedly in her *Face of the Deep* as the creative *chora*, that birthed Carson's awareness of our embededdness in systems that are larger and more than us. "Carson portrays the ocean as vast, indomitable, and omnipotent. Humans, by contrast are depicted as small and somewhat insignificant, as unable to control and master the ocean's power and to harness it for our own ends."[44] Though she does not develop a full-blown apophatic epistemology, it can be argued that her relationship to and reflections on the sea have several components of what we might call an art of unknowing and an acceptance that nature lies beyond the capacity of human, objective reason to exhaust. Two components of such an epistemology that I want to highlight here are her critique of human mastery and the importance of fostering a sense of wonder. Neither of these suggests that she was not adamant about gathering scientific data: she was, after all, a scientist. However, as with Haeckel, her scientific inquiry was couched in a wider sense of mystery and wonder, and this is key to her appreciation for the rest of the natural world. There were, of course, many other scientists—including Eugene Odum and Aldo Leopold for instance—who were also concerned about nature, but mostly their versions of ecological understanding were focused on metaphors of management and stewardship. Carson introduced a multiagential, and maybe even multinatural, understanding of human-earth relations in which we humans are not fully in charge, harkening back to the "nature" of Goethe, von Humboldt, and Haeckel.

Carson, though agnostic and raised in a Presbyterian family, might today be what we call a religious naturalist or what Nassar calls a "romantic empiricist." In other words, the mystery of the natural world provided for her a wider sense of value and purpose in life. She was influenced by Albert Schweitzer's "reverence for life," and such reverence does show up in her delicate balance between objective, scientific observations and respect for the mysteries of the rest of the natural world. We may look back on her life now and see her in a single narrative line as one of the mothers of the modern environmental movement, but the idea that the chemicals we used in agriculture and other parts of our daily lives could drastically change our environment and have repercussions for our own health was not of wide concern. She writes in the era of the Green Revolution, after all, and technology and science were supposed to save us from the lack of natural provisions. Furthermore, the idea that we might cause enough harm to our fisheries and other ecosystems to the point of their collapse was not one that was widely held or even thought about. Finally, she was already thinking about climate change and the melting of glaciers in the 1960s and thus beginning to realize the planetary interconnectedness of humans and nature.[45] All of these connections were not made by the dominant reductive and productive scientists. Thus, her work entailed bucking the dominant ethos of the scientific community as well as the wider society. Much like Barbara McClintock and her genetic discoveries based upon a more holistic (and less reductive) understanding of corn plants, so Rachel Carson's breakthrough came in part through her deep listening and humility.[46] Her work entailed a great deal of unknowing of received knowledge policed by the institutions of mechanistic science. Just because she was critical of the mechanistic view did not mean, of course, that she completely embraced some sort of a feminist or ecofeminist organic view. Rather, she blurred the boundaries between these two models.

> As Carson incorporated the human body into "ecology," she also began to resolve a longstanding tension between two distinct narrative voices with which she had experimented in her major works on the sea: the "heavily masculine" voice of the scientist who coolly organizes, interprets, and summarizes reams of information, and the more feminine voice of the "appreciative nature writer," the close observer and participant, expressing wonder and enchantment with nature's mysteries.[47]

In a way we might say that Carson participated in a queering of the perspective of the relationship between humans and nature, and in doing so challenged the gendered stereotypes assumed by that distinction. I'm not suggesting here that Carson was any sort of queer theorist, but rather that she destabilized the ways in which gendered views of nature were used to anchor gender norms simply by destabilizing human-earth relationships. In doing so, she challenged both the way in which established science "knew" nature and what "proper" male and female outlooks on the world *are*. She also paid heavily for these challenges as established scientists (and the agro-industry funding many of them) fought her in the media, attacking her intellect and personal character with great force, and always with the twinge of questioning her gender and sexual orientation in the background.

Again, the method by which she comes to this blurring of boundaries is, perhaps not insignificantly, best articulated in her "sea ethic." Carson sought to understand the life of the sea from the perspective of organisms that inhabited the sea: in the coastal areas, the shallow waters, and even the "abyss." Exploring the life of the deep sea was very much a new area at the time when she was writing about the sea. Accordingly, she attempted in her writing and research to take on multiple perspectives of sea life and resisted the totalizing tendency in scientific knowledge to explain everything as fact. One might say that she allowed the contours of the ocean to change the way that she thought about scientific observations. She did so not in a naive way that confused her human perspective with the ocean itself, but rather in a critical way. She developed, in other words, what my colleague Lisa Stenmark, among others, has described as "an ecology of knowledge." A type of knowledge that takes on the contours and engagements of what the observer is immersed in; it is precisely not a perspective from nowhere, but one that is embedded and context dependent. A couple of examples might suffice. First, Carson writes,

> It is not the distance from the shore, but the depth, that marks the transition to the true sea; for wherever the gently sloping sea bottom feels the weight of a hundred fathoms of water above it, suddenly it begins to fall sway in escarpments and steep palisades, descending abruptly from twilight into darkness.[48]

This is a typical statement from Carson: part scientific observation, part wonder, and part recognition of the limits of human knowing. Such limits provide the habitat for new insights "from the deep" to emerge.[49] It provides the space for otherness so that we don't pretend that through an expansion of our ego's we can "think like a mountain." This is essential for a CPR. In yet another blurring of boundaries, she articulates a type of deep understanding of the interconnectedness of all life and of the multiplicity of subjectivity. While observing the life cycle of the mackerel, she writes,

> Of the millions of mackerel eggs drifting along . . . thousands went no farther than the first stages of the journey into life until they were seized and eaten by the comb jellies, to be speedily converted into the watery tissue of their foes and in this reincarnation to roam the sea, preying on their own kind.[50]

Haeckel's argument that in the end everything is transformed sunlight couldn't partner any better with Carson's thinking here. This observation is indicative, again, of the way in which she sees life as interconnected and in the way in which she understands the mystery/wonder of life. It is not too much of a stretch, then, to understand Carson in a long line of scientists who were rethinking nature from a nonreductive, relational-based understanding of reality in contrast to a reductive, mechanical one. Such a multinatural and multiperspectival metaphysic would fly in the face of a certain type of monotheistic/monological understanding of nature, humans, and the world in which knowledge and reality were in a fixed and knowable relationship. Again, this is one reason her work was so threatening for so many.

Her work continued to unravel the frayed edges of a metaphysical system that ordered the Western relationship between humans, knowledge, and the natural world in a dualistic and hierarchical way. It also flies in the face of certain postmodern ideas that suggest humans fully construct nature, as if humans are somehow the only agents in the world and nature, even in the "post"-world, is still dead stuff. This backgrounding of nature, or fear of the natural, has been the Achilles's heel of postmodernity.[51] It maintained some of the most violent constructed realities as real: that humans are somehow above the rest of the natural world, that humans are the only agents in the world, and that nature can be molded to meet human desires and needs. Carson, no postmodern thinker, challenges these latter points and helps us understand how much we are a part of nature naturing, a part of the wider planetary community. This is important. Whereas Haeckel had the Romantic connection, it was not planetary in scope and humans were still at the apex of evolution. This meant that getting rid of God and replacing it with Nature, only left Haeckel with his culturally-socially-biologically located understanding of nature, which he mistook for Nature itself. From this perspective, everything had to fit into Haeckel's view of nature. Sadly, the predominant events shaping the worldview of Haeckel's second half of life were nationalism and a world war. He didn't have a strong sense of the planetary, and thus the planet was fitted (again) into his own human making, and various life-forms were ranked accordingly. Just as he critiqued the theologians for projecting their own understanding of the human onto an omni-god, so he projected his own understanding of human reality onto Nature. Scientists such as Rachel Carson added a more critical form of Romanticism and a planetary perspective to evolutionary and ecological thinking, but not without controversy. Carson's scale of concern was the planetary, and the planetary is what I am arguing is Latour's "critical zone" within which we need to think, hope, breathe and live. Not the cosmos, not the local or national, but the planetary. From this space we might begin to think of a planetary polis.

Carson's witness and more importantly her humility have literally transformed the world in ways that allow for many more planetary voices and connections to come into speech and for new technologies to emerge that take those planetary voices into account and thereby help to co-create natures in a different way. Carson would end up, tragically, giving her life as part of the proof that chemicals in our environments can lead to "wicked problems" for both humans and the rest of the natural world. Though one might argue there is no direct causality between agro-chemicals and breast cancer, few would argue the evidence of the strong positive correlations. The revelations Carson made through her work have been crucial for the unfolding of the environmental justice movement.

In Carson's example of a type of unknowing that leads to new ways of knowing, I think we find a good model for a planetary, embodied way of understanding how we claim to know what we know about the world. It is not as easy as applying past knowledge on to the future becoming of life and thereby forcing a repeated sameness onto future generations. This is the quick and easy form of reasoning and working through problems: We make solutions based upon what we know. This is what modernity did with certainty, and what postmodernity did with maintaining that "nature" is constructed largely by humans and culture. That latter certainty, or hubris, has helped maintain a fracture between progressive social politics and environmentalisms even to this day.

What I would argue, in a sense, is that Rachel Carson, like Romantic empiricists before her, was not only shifting our understandings of human beings, but also of nature in ways that suggest a sort of multinaturalism that meets the multiculturalism of the postmodern twentieth century. It is not just that culture describes in various ways what a single natural world means, but that our understandings of and interactions with nature from various positionalities help to co-create different actual and possible worlds for planetary becoming. We experience different natures given our embodiments and locations within the worlds that make up our planet. We should, then, be wary of any system of thought (religious, scientific, philosophical, political, or some combination thereof) that attempts to reduce the multiplicity of the planet into a single narrative. We must always, rather, think in terms of open and multiple narratives.

CRITICAL PLANETARY ROMANTICISM AND THE GROUNDING OF HOPE

Throughout this book, I have been engaged in a constructive retrieval of sources for a nonreductive naturalism, new materialism, "Romantic empiricism" or what I am calling a CPR. In retrieving some sources for this nonreductive naturalism, I have argued that placing humans within a naturalistic and evolutionary perspective means also that we must let go of the certainty provided by transcendent objectivity, by eternal essences and forms, and by teleology. These types of certainty, I argue, lead to dangerous projections onto the rest of the natural world which, when taken as natural rather than as co-constructed interpretations, can lead to the justification of hierarchies of oppression: They abstract us from the world and hinder our ability to reattune to evolving contexts. Instead, these abstractions try to mold the world into their own images. These are the problems that some of Haeckel's ideas fell victim to precisely because he himself was certain of one type of truth or another. It is my argument that in a postindustrial globalized world for some, in which others are being incentivized and encouraged to industrialize in the same way as Western moderns, and in a world marked by climate change, we can re-engage these earlier Romanticisms without falling victim to the same types of certainties many earlier scientists and thinkers did because we now have the resources to think of ourselves as planetary creatures. We can be even more critical today than a hundred years ago thanks to the growing, pluralistic understandings of nature-cultures. We can acknowledge multiple perspectives about the world, none of which provide objective, bare facts about the world. Furthermore, we can assert a type of multinaturalism that acknowledges that nature changes along with the relationships that make up nature over time, and that the various dynamic and diverse ecosystems that make up the planet at any given time require a multinaturalistic outlook.

Finally, I have argued that the idea of the death of nature beginning with the so-called Western Scientific Revolution and challenged in the twentieth century with the rise of Western environmentalism is too simplistic of a narrative to account for the many diverse methods and understandings of the rest of the natural world. Whereas the "dead matter" of a reductive and productive understanding of the natural world may be true of a certain interpretation of Newtonian physics, it was not so for most of the biological sciences of the nineteenth century. Many Romantic empiricists were voicing nonreductive naturalisms. It was only amid the horrors of World War II that the natural and earth sciences became reduced to the reductive and productive model of science (though never completely). Just as theoretical physics was beginning to be challenged by Einsteinian and post-Einsteinian physics, the natural and social sciences, and even the humanities, began conforming to a reductive, productive, and discipline-based model. Sadly, this reductive/productive mentality still has hold of the curricula in most U.S. universities (among others) today. Though never fully embraced, the reductive productive model reigned supreme for about twenty to twenty-five years until the environmental, civil rights, and liberation movements began challenging reductive models of the social sciences and humanities.

What these reductive models all have in common (whether *Naturwissenschaften* or *Geisteswissenschaften*) is an attachment to certainty from either the past (in the form of essences and forms), the present (in the form of an objective human standpoint), or the future (in the form of teleology). As I have argued throughout this text, placing humans within an evolving, planetary context might do away with these certainties and allow space for a CPR that relies on a bit of unknowing at the edges of our worldings, and constant critical reattunements to the new worlds in which we live.

Here, at the end of this book, it is only appropriate that I focus on the future. Whereas models of justification based upon some form of certainty close off worlds, a CPR for the earth understands that the processes of planetary becoming are open to an unknown future. Legitimization of knowledge claims will always be open-ended and decided on from future perspectives.

This is a form of what might be called "horizontal transcendence," which relies on the fact that the future becoming of all life is open-ended.[52]

Like so many other things within human histories and cultures and within individual lives, salvation or condemnation comes from beyond. Not necessarily an invisible, supernatural beyond, but from the unknown or what Ernst Bloch called the "not yet."[53] Buddhism's focus on the present, on "is-ness" also captures the unknown, not-yet of the future, as do many other understandings of time in various religious traditions. Moments of creative insight, moments of clarity, moments of innovation and change, moments of great love for one another, as well as moments of fear and trepidation, often come from beyond to lure us into new ways of thinking, relating, and becoming. This "delayed messianism" (Derrida) may be at the heart of the meaning of the incarnation of hope in the world: To create a different possible planetary community, we need something from beyond to help us shift toward new ways of becoming. Humans seem to be incapable of saving themselves and the planet.

From within this understanding of the justification for actions and knowledge as "to come," we might again begin to live in hope, with(in) the planetary community, and work together toward planetary futures that are better able to accommodate for the flourishing of various bodies and perspectives that make up the communities of the planet at any given time. In other words, we might remain open to one another within the planetary community in order that we might learn how different ideas, knowledge regimes, and imaginings materialize in the world to affect planetary bodies in various ways, for better and worse. We must constantly reattune to the worlds in which we live as these worlds continue to shift. As we listen to one another, we can correct for the violence enacted upon some planetary bodies and rethink again what it is we ought to do. As Hannah Arendt was so well aware of, we cannot predict how our actions will ripple out through space and time to affect various bodies differently, yet we must act.[54] The best we can do, then, is to adopt an experimental approach toward knowledge and value claims, one version of which I am calling a CPR for the earth, that recognizes the "not yet" of justification, which flows from not-yet existing future possibilities of planetary becoming.

AFTERWORD

Critically Navigating the Planetary Seas

As I have been writing and talking more and more about things like uncertainty, multiplicity, undisciplined thought, and developing a CPR, usually the first question I get is about how to develop an ethics or politics out of such nonfoundational, unstable thinking. Similarly, if identities are co-constructed over time and all life is in some way queer, how do we call for justice for specific embodiments that have been systematically oppressed due to their race, sex, sexuality, ability, or gender? There are also the usual questions that emerge out of projects attempting to (re)place humans within the rest of the evolving planetary community as part of that community rather than masters of the entire planet. For example, how do you decide between who lives and dies, or what constitutes a "just" or "good" amount of death since all life lives at the expense of other lives? I don't pretend to have good answers or hard and fast guides to these moral and ethical questions. However, in an era of "fake news" and "alternative facts," I realize that I have a responsibility to provide some sort of compass, some sort of guidelines for navigating between competing claims of truth, goodness, and beauty.

As I was finishing the first draft of this book, in February 2023, I had the pleasure of hosting Vandana Shiva at Florida International University. In speaking with her about this book, and my work in general, she pushed me to think about what markers for "better" vs. "worse" planetary ideas and thinking might look like. This resulted in the following four guides for good, critical planetary thinking. First and foremost, before even getting to the guides, we need to figure out how to think together in a planetary polis, or "earth democracy," or as "terrapolitan" citizens.[1] This is no small order given the current (as of 2026) rise of nationalisms and localisms across the globe that would have us retreat to geographic, cultural, racial and other "local" borders, and given the outright attack on institutions of higher education in the United States and elsewhere. The erosion of public spaces (in the Arendtian sense of a space of political dialogue) has now unfortunately reached curricula and classrooms.[2]

Anecdotally, as of this writing I have been teaching at a public university in the state of Florida for about seventeen years. The university is very diverse: About 70 percent of the students are first generation to college and about the same are ESL students (meaning they don't speak English in the home as a first language). Each classroom is in many ways a microcosm of the planetary. When I first arrived in 2008, I found it much easier to engage my diverse students in group and class discussion around thorny political and ethical issues. Today (2026 at the time of this writing), I find that students are less willing to have an open dialogue in the classroom. This is in part due the hostile takeover of higher education by the Trump administration and his followers, who seek to end the "woke indoctrination" of students and replace it with "classical learning." Dismantling critical thinking, suing universities, firing faculty, instituting textbook review committees, terrorizing all sorts of immigrants through ICE, and purging institutions of funding and courses that don't agree with the prerogatives of the Trump administration is all too similar to the rise of fascism in Europe right around the time of Haeckel's death. Of course, the Right is "grooming" just as much as they accuse the Left of doing, but they seem to have the political power now to enforce their own ideology across the nation (and elsewhere). How then do we even begin to navigate between indoctrination and critical thinking? What does all of this mean for learning and education when students have lost the ability and courage to speak freely and openly with one another? What does it mean for local, regional and planetary politics? Furthermore, what can be done in this situation of noncommunication other than battening down the hatches and preparing for ideological warfare? It seems to me that we (and here I mean educators, but also leaders of all sorts) need to pause and think about what knowledge is, and what it is for, before we can answer the question of how we might move forward. In short, I see two broad systems of knowledge production that provide us with different ethical ends: one that I call reductive, and the other I call connective.

The reductive model of knowledge, a revival of the "Mint Julep" model,[3] is one that reduces knowledge to single perspectives, truths, and approaches. You can find this model in the dogmatism of the far right and the far left. You can find it in the fundamentalism of religious beliefs. You can also find it in some versions of scientific reductionism: it's all in the genes, or it's all chemical, or it's all in the neurons, or more broadly, the idea that science is all there is and all that matters. This model tends to place (some) humans as "exceptional" to the rest of the natural world and as somehow capable of having an objective, bird's-eye view of the rest of the natural world, which can then be explained to everyone else. This view, because it argues for a single perspective/truth, also places a single human perspective as *the* correct perspective. There are, in other words, no critical approaches to knowledge, but there is just correct knowledge and everyone must assent to or recognize it. It is this type of reductive knowledge that eliminates all other perspectives than "the one" (which usually belongs to the one staring back at you in the mirror).

The connective model of knowledge operates from a different starting point. It is a model of what Freya Matthews calls encounter. She writes that "the materialist [what I am calling the reductive view] view of the world that is a corollary of dualism maroons the epistemic subject in the small if charmed circle of its own subjectivity, and that it is only the reanimation of matter itself that enables the subject to reconnect with reality."[4] We humans are a part of an evolving community of life, and there is no outside space from which to gain an objective, bird's-eye view of the world. In fact, no such space exists: there are just multiple perspectives of the common planet on which we live, and which we co-constitute and are co-constituted by. This is not about relativity or an "anything goes" mentality; rather, it is about embodiment and context: If we emerge from an immanent process of planetary becoming, then each embodiment/entity within that planetary becoming has something unique to tell us about the makeup of the planetary community at any given time. From the perspective of this type of knowledge, the gauge of whether or not knowledge is "good" knowledge, and whether or not an action is "good," is how much it helps us to think with others (both human and nonhuman) and work toward co-constructing worlds that provide political spaces for multiple perspectives and ways of being/becoming to thrive. Certainty and single-mindedness are the enemies of this system of knowledge production.

As the reader of this book is likely aware, I am partial to the connective model, and for good reasons. I think this type of knowledge creates more connection between the *humus* of humanity and humility, and the *humus* that is the dirt of the earth. It goes a long way to prevent knowledge (and some humans) from fleeing the planet and its many creatures. This connective form of knowledge also helps to foster compassion: We can identify the ways in which multiple others are suffering as a result of long histories of colonization, human exceptionalism, sexism, racism, and genocide. With knowledge focused on humility and compassion, we might build genuine political spaces of conversation: spaces in which we enter dialogue with the genuine possibility of being converted by other perspectives. Finally, from these spaces of political dialogue and conversation (again in the Arendtian sense) we might find new companions: companions which we not only share bread with, but more broadly food and all other earthly resources. Here, at the end of this manuscript, let me offer these guides for critically, planetarily, and romantically reattuning to the worlds around us.

THE HUMILITY OF *HUMUS*

The etymological connection between *human*, *humility*, and *humus* (earth) is one that, I think, cannot be highlighted enough. There are, of course, many origin stories that have humans emerging from the dirt: Adam from the Adamah in the Torah, Quran, and Christian Bible; the fashioning of humans from clay in Greek, Egyptian, and Chinese origin stories; the many Indigenous stories that understand humans as being created from the clay or earth and that humans are "kin" with other life on the planet; and, of course, the scientific story of the evolution of life is one that suggests that life emerges from the oceans and then evolves out of the earth. The point is that the human is deeply tied to the earth in many religious, cultural, and scientific understandings of the origins of humanity. This is the first recognition needed for reattuning to the evolving planetary communities.

Rather than act as if we are not of the *humus*, humility helps us to recognize that we are deeply connected to the dirt, the earth, and all life therein. Those knowledge systems and actions that promote this connection and deepen it, then, might lead to a "better" planetary future. In other words, that which promotes our connection to the earth also promotes planetary truth, goodness, and beauty.

THE OPENNESS OF COMPASSION

The knowledge of our entanglement with the *humus*, though not always a "good" or "beautiful" thing for the individual, may help open us toward more compassion. I would argue that there is some truth to the Buddhist understanding that life is suffering: everything lives at the expense of other life; our entanglement with the rest of the natural world means that we can be infected by viruses and bacteria, and that eventually our bodies will become food for other bodies. Our entanglement means that our relationships with humans and the rest of the natural world can harm and hurt, and that other humans and the rest of the natural world can be quite indifferent to our own needs. The etymology of the word compassion is to "suffer with." And this is the second recognition needed for reattuning to the evolving planetary community around us.

I am not suggesting that we all become martyrs or suffering servants. Rather, I am saying that knowledge which opens our eyes to the suffering of other lives and opens our hearts to the pain of this suffering is knowledge that is good, true, and beautiful for the planetary community. This is not just the attention that the joys and pains of our daily lives call for, but also attention to the suffering caused by racism, sexism, ableism, heteronormativity, anthropocentrism, and many other institutionalized forms of violence in our worlds. The suffering caused by fossil-fueled climate change, by colonialism, and by hyperextractive consumerism caused largely by the globalization of neoliberal economic systems should not be something that we just think of as "life as usual." "To suffer with" means to try to understand, and to stand with all those in the planetary community who are suffering in multiple ways, with an eye toward both witnessing that suffering and, when possible, alleviating that suffering by working toward worlds that are more truthful, good, and beautiful.

THE PLAYFULNESS OF CONVERSATION

I think the prerequisites for developing something like a planetary polis, are the types of humility and compassion I have articulated above. What does it look like to have a real conversation? The etymology of conversation is something like "to turn together." From my interpretation this means the ability to enter into dialogues with a willingness to be converted to other's viewpoints. This type of conversation is, again, at the heart of Arendt's understanding of the political sphere. However, I would extend that sphere (which she does not) to the rest of the planetary community, the rest of the natural world. How might we be converted to ways of becoming that we learn from being in conversation with dolphins, whales, birds, primates, elephants, and other forms of animal and plant life? How might we hear the cries of species extinction, warming oceans, deforestation, and habitat loss in our efforts to "turn together" toward something "better"?

The ability to open oneself up to possible conversions in conversation with the entire planetary community is also a marker of what is true, beautiful, and good. Conversation stoppers and ideologies that do not entertain the knowledge and ideas of others are most often harmful. These open conversations are hard work, but there is also a sense of playfulness we might cultivate that helps, in the words of Nelle Morton, "hear others into speech."[5] The playfulness of song, poetry, creative non-fiction, speculative fiction, theater, and art (both human and non) may help us cultivate spaces that foster productive conversations: and these types of spaces are the spaces required for a planetary polis.

THE JUSTICE OF COMPANY

Finally with humility, compassion, and conversation, we might begin to develop more just ways of being together. The etymology of *company*, "with bread," is simply the tradition of hospitality known as breaking bread together or eating a meal together. There is perhaps nothing more basic that humans (and nonhuman others) can do together. Of course, this is not always done in a just manner: some are impoverished while others indulge; the violence of factory farms and monocultures dependent upon pesticides and fertilizers and fossil-fueled energy in general cause much suffering and environmental destruction; the labor practices of those who enable a meal to be on a given table are unequal and often times violent (depending on where one is). If we can approach our company with humility, compassion, and in a spirit of conversation, we might be able to better address these issues of justice (for humans, animals, and the rest of the natural world).

The more we create spaces for just communities, the more true, good and beautiful the planetary community might become. This means creating spaces, and the infrastructures, policies, and institutions within these spaces, that address all the "isms" and the violence caused by those isms. Company that is not critical of its own terms of existence can serve to perpetuate all the social and ecological violence that a critical planetary romanticism hopes to confront and diminish.

At the end of this lengthy, theory-laden book, falling back on humility, compassion, conversation, and company might seem trite or unsatisfactory. But each of these guides for bringing more goodness, truth, and beauty into the world are context-dependent and shifting, just like the needs and desires of the planetary community. Those forms of knowledge and truths which increase these four emergent and evolving ideals, then, may serve as better guides than any certain axiom, revelation, law, intuition, or universal imperative ever will.

NOTES

INTRODUCTION: RETHINKING NATURE, RELIGIOUSLY—REATTUNING TO OUR WORLDS

The epigraphs are taken from Ernst Haeckel, *The Riddle of the Universe at the Close of the Nineteenth Century* (New York: Harper, 1900), 19, 20.

1. Val Plumwood, *Environmental Culture: The Ecological Crisis of Reason* (New York: Routledge, 2002), 27–30.
2. Philippe Descola, *Beyond Nature and Culture* (Chicago: University of Chicago Press, 2013).
3. Dalia Nassar, *Romantic Empiricism: Nature, Art and Ecology from Herder to Humboldt* (New York: Oxford University Press, 2022). Though not necessarily aligned with the typical period known as the "Romantic era," *Romanticism* is assumed in this book to begin with that period of reawakening to the natural world during the Industrial Revolution and extend up to the present.
4. See, e.g., Whitney Bauman, *Religion and Ecology: Developing a Planetary Ethic* (New York: Columbia University Press, 2014); and Laura Jean Vollmer, "The Relationality of Religion and Science: Toward a New Discourse-Analytical Framework" (PhD Dissertation, University of Groningen, 2017).
5. Kate Rigby, "Nature, Language, and Religion: Herder and Beyond" in *Ecological Thought in German Literature and Culture*, ed. G. Duerbeck, U. Stobbe, H. Zapf, and E. Zemanek (Lanham, MD: Lexington Books, 2017), 31–41.

6. Brycchan Carey, *The Unnatural Slave Trade: Slavery, Abolition and Environmental Writing, 1650–1807* (New Haven, CT: Yale University Press, 2024), 103–105. As Carey argues in this book, the clerical naturalists were also embroiled in a debate about whether slavery was natural or unnatural based upon the close study of creation, something that would haunt later nineteenth- and twentieth-century science as well in terms of the issue of nature and race.
7. Freya Matthews, *For the Love of Matter: A Contemporary Panpsychism* (Albany: Statue University of New York Press, 2003), 88.
8. J. R. McNeill, *The Great Acceleration: An Environmental History of the Anthropocene Since 1945* (Cambridge, MA: Harvard University Press, 2014).
9. Terra Schwerin Rowe, *Of Modern Extraction: Experiments in Critical Petro-Theology* (London: T&T Clark, 2022).
10. Mel Chen, *Intoxicated: Race, Disability, and Chemical Intimacy Across Empire* (Durham, NC: Duke University Press, 2023), chap. 1.
11. Michael Zimmerman, *Heidegger's Confrontation with Modernity: Technology, Politics and Art* (Bloomington: Indiana University Press, 1990).
12. A good analysis of this sped-up process of globalization can be found in Teresa Brennan, *Globalization and Its Terrors* (New York: Routledge, 2003).
13. Whitney A. Bauman, "Prismatic Identities in a Planetary Context," in *Ecological Solidarities: Mobilizing Faith and Justice for an Entangled World*, ed. Dhawn Martin, Elaine Padilla, and Krista Hughes (University Park: Penn State University Press, 2019), 186–200
14. Timothy Morton, *Hyperobjects: Philosophy and Ecology After the End of the World* (Minneapolis: University of Minnesota Press, 2013).
15. Rob Nixon, *Slow Violence and the Environmentalism of the Poor* (Cambridge, MA: Harvard University Press, 2013).
16. Thomas Princen, Jack P. Manno, and Pamela L. Martin, eds., *Ending the Fossil Fuel Era* (Cambridge, MA: MIT Press, 2015).
17. Lisa Sideris, *Consecrating Science: Wonder, Knowledge, and the Natural World* (Berkeley: University of California Press, 2017).
18. Donna Haraway, *Staying with the Trouble: Making Kin in the Chthulucene* (Durham, NC: Duke University Press, 2016).
19. Haraway, *Staying with the Trouble*, 65.

20. Carolyn Merchant, *The Death of Nature: Women, Ecology and the Scientific Revolution* (New York: HarperCollins, 1980).
21. Anna Peterson, *Being Human: Ethics, Environment and Our Place in the World* (Berkeley: University of California Press, 2001).
22. Carolyn Merchant, *Reinventing Eden: The Fate of Nature in Western Culture* (New York: Routledge, 2003); Carolyn Merchant, *Autonomous Nature: Problems of Prediction and Control from Ancient Times to the Scientific Revolution* (New York: Routledge, 2016).
23. For instance, Terra Schwerin Rowe shows well how the excitement around the extraction of oil in the nineteenth and early twentieth centuries was not based upon humans' use of dead matter, but rather contained quite a bit of excitement around the animacy and animating powers of oil. See her *Of Modern Extraction*.
24. The internalization of wonder, and wonder at human knowledge gained, is made possible, for example, through the "death of nature," and the subsequent understanding of the human as the only agent that matters, during the Anthropocene. I'm building here on Lisa Sideris, *Consecrating Science*, 22–26.
25. Walter Mignolo, *The Darker Side of Western Modernity: Global Futures: Decolonial Options* (Durham, NC: Duke University Press, 2011).
26. For this reason, Kevin Minister argues that we don't need to think about the future in terms of "sustainability," or more of the same projected out, but rather apocalypse: the end of a world and beginning of a new one. See Kevin Minister, "Interreligious Approaches to Sustainability Without a Future: Two New Materialist Proposals for Religion and Ecology," in *Earthly Things: Immanence, New Materialisms, and Planetary Thinking*, ed. Karen Bray, Heather Eaton and Whitney Bauman (New York: Fordham University Press, 2023).
27. Bruno Latour, *If We Lose the Earth, We Lose Our Souls* (Cambridge: Polity Press, 2024).
28. Latour, *If We Lose the Earth*, 15.
29. The rhizome of Deleuze and Guattari is also relevant here: Gilles Deleuze and Felix Guattari, *A Thousand Plateaus: Capitalism and Schizophrenia* (Minneapolis: University of Minnesota Press, 1987); Isabelle Stengers's metaphor of the palaver is also relevant here: Isabel Stengers, "The Cosmopolitical Proposal," in *Making Things Public*,

ed. Bruno Latour and Peter Weibel (Cambridge, MA: MIT Press, 2005), 994–1003.

30. Or, with Isabele Stengers, we might think of the palaver: "Cosmopolitical Proposal," 1003.
31. Mignolo, *Darker Side of Western Modernity*, 71–72.
32. See, e.g., Mario A. Di Gregorio, *From Here to Eternity: Ernst Haeckel and Scientific Faith* (Göttingen, Germany: Vandenhoeck and Ruprecht, 2005), 27; and Ernst Haeckel, *Gott-Natur (Theophysis)*, mit *Kommenter Nachdruck von Olaf Breidbach und Uwe Hoßfeld* (Stuttgart, Germany: Franz Steiner Verlag, 2008), 18.
33. Rigby, "Nature, Language and Religion."
34. Friedrich Schleiermacher, *The Christian Faith* (London: T&T Clark, 2016), 12–13.
35. David Strauss, *Das Leben Jesu: Für das Deutsche Volk Bearbeitet* (Leipzig, Germany: F. A. Brockhaus, 1864).
36. Ludwig Feuerbach, *The Essence of Christianity* (London: Kegan Paul, Trench, and Trübner, 1890).
37. This division between the humanities and natural sciences is also quite problematic, and more will be said about this in later chapters. Sylvia Wynter, for instance, argues that this basic division in modern Western education gone global is itself at the root of the colonial mentality for holding on to a specific ideal of the human, and for separating humans out from the rest of the natural world. See Sylvia Wynter, *On Being Human as Praxis* (Durham, NC: Duke University Press, 2015).
38. Dalia Nasar, *The Romantic Absolute: Being and Knowing in Early German Romantic Philosophy, 1795–1804* (Chicago: University of Chicago Press, 2014).
39. And, not inconsequential, the Romantics were heavily influenced by the more relational metaphysics found in "Eastern" philosophy and thought such as Buddhism and Hinduism, and from Romantic thinking in the Muslim world.
40. Jeffrey Einboden, *Islam and Romanticism: Muslim Currents from Goethe to Emerson* (London: One World Publications, 2014), 14.
41. Donna Haraway, *When Species Meet (Posthumanities)* (Minneapolis: University of Minnesota Press, 2008).

42. See H. Hohlwein "Keplerbund," in *Die Religion in Geschichte und Gegenwart*, Band 3, (Tübingen, Germany: JCB Mohr, 1959); and Weber and Breidbach, "Der Deutsche Monistbund," 174.
43. Robert J. Richards, "The German Reception of Darwin's Theory, 1860–1945," in *The Cambridge Encyclopedia of Darwin and Evolutionary Thought* (Cambridge: Cambridge University Press, 2013), 240.
44. Erika Ellis, "The Reception of Ernst Haeckel in America: Monism and The Prophet of Evolution" (undergraduate thesis, University of Washington, Seattle, 1998).; see also Ernst Haeckel, *Gott-Natur (Theophysis), Kommenter Nachdruck von Olaf Breidbach und Uwe Hoßfeld* (Stuttgart, Germany: Franz Steiner Verlag, 2008), 25: "Affenbrief von an anonymous American: 'Gott hat bei der Schöpfung nicht gesagt, lasset uns Affen machen, sondern lasset uns Menschen machen, ein Bild das uns gleich sei. So viel ich vernommen habe, willst Du auch in Amerika Vorträge halten, das kannst Du sein lassen, hier hat es Affen genug von Afrika, wir brauchen daher keine deutschen Affen mehr.'" Finally, see Weber and Breidbach, "Der Deutsche Monistbund," 174: "Gerade die neuen Entdeckungen der Relativitäts- und Quantentheorie dienten dem Keplerbund als Nachweis für die Relativität wissenschaftlicher Erkenntnisse und als schlagender Beweis gegen den ontologischen Dogmatismus der Monisten."
45. Depending on where you search, the results are different, but they usually include at least one link such as the following, which was searched from Jena, Germany, and is from Creation Ministries International: Russel Grigg, "Ernst Haeckel, Evangelist for Evolution and Apostle of Deceit," Creation Ministries International, January 27, 2006, updated February 4, 2015, http://creation.com/ernst-haeckel-evangelist-for-evolution-and-apostle-of-deceit (first published in *Creation* 18, no. 2 [1996]: 33–36).
46. Jonathan Marks, "Why Were the First Anthropologists Creationists?" *Evolutionary Anthropology* 19 (2010): 222–226. On p. 224, speaking of Virchow, he writes, "Regardless of whether it could be established that this point actually motivated these scholars against Haeckel's views, it is nevertheless true that they did oppose Haeckel's views, that Haeckel's views did fundamentally undermine their intellectual program by implicitly denying the psychic unity of mankind."

47. For instance, he argues that Kant's distinction of knowledge as "empirical, or a posteriori—that is, derived from experience; and that the rest of our knowledge (as, for instance, mathematical axioms) is a priori—that is to say, reached by the deductions of pure reason, independently of experience. This error led to the further statement that the foundations of science are metaphysical, and that, though man can attain a certain knowledge of phenomena by the innate forms of space and time, he cannot grasp the "thing in itself" that lies behind them"; Ernst Haeckel, *The Wonders of Life: A Popular Study of Biological Philosophy* (New York: Harper, 1904), 10. In a letter to Wilhelm Bölsche, Haeckel signs his name in a way that suggests the move to monism from theology is: "From Plato to Darwin :: From Kant to Haeckel" (February 2, 1905), found in Rosemarie Nötlich, *Ernst Haeckel-Wilhelm Bölche: Briefwechsel 1887–1919* (Berlin: Verlag für Wissenschaft und Bildung, 2002), 171.
48. Catherine Keller, *Cloud of the Impossible: Negative Theology and Planetary Entanglement* (New York: Columbia University Press, 2014); Karen Bray, Heather Eaton, and Whitney Bauman, eds., *Earthly Things: Immanence, New Materialisms, and Planetary Thinking* (New York: Fordham University Press, 2023).
49. See, e.g., Jacques Derrida, *Of Grammatology* (Baltimore: John Hopkins University Press, 1974); and Judith Butler, *Bodies That Matter: On the Discursive Limits of Sex* (New York: Routledge, 1993).
50. Haeckel, *Wonders of Life*, 10–11.
51. Michel Foucault, *The Order of Things: An Archaeology of the Human Sciences* (New York: Random House, 1970).
52. Donovan Schaefer, "The Things of Order: Affect, Material Culture, *Dispositif*" in *Cultural Critique* 124 (Summer 2024): 16–17.
53. On Kant's false move to the a priori/ideal in his epistemology, Haeckel writes, "This dogma is erroneously built on the correct idea that our knowledge, obtained through the senses, is imperfect; it extends only so far as the specific energy of the senses and the structure of the phronema admit. But it by no means follows that it is a mere illusion, and least of all that the external world exists only in our ideas"; Haeckel, *Wonders of Life*, 69.
54. Karen Barad, *Meeting the Universe Halfway: Quantum Physics and the Entanglement of Matter and Meaning* (Durham, NC: Duke University Press, 2007).

55. Ernst Haeckel, *The Riddle of the Universe at the Close of the Nineteenth Century* (New York: Harper, 1900), 91: "And we must say just the same of the higher psychic activity of the higher animals and man, of the formation of ideas and concepts, of the marvellous phenomena of reason and consciousness; for the latter have been phylogenetically evolved from the former, and it is merely a higher degree of integration or centralization, of association or combination of functions which were formerly isolated, that has elevated them in this manner."
56. Marsha Morton, "From Monera to Man: Ernst Haeckel, Darwinismus, and Nineteenth Century German Art," in *The Art of Evolution: Darwin, Darwinismus, and Visual Culture* ed. Barbara Larson and Fae Brauer (Hanover, NH: Dartmouth College Press, 2009), 63.
57. Ernst Haeckel, *History of Creation: Or the Development of the Earth and Its Inhabitants by the Action of Natural Causes*, vol. 1 (New York: D. Appleton, 1876), chap. 15.
58. Ernst Haeckel, *Kristallseelen: Studien über das anorganische Leben* (Leipzig, Germany: Alfred Kroner Verlag, 1917).
59. Alfred North Whitehead, *Process and Reality* (New York: Free Press, 1978), 83–109.
60. On Whitehead and attunement, see Alfred North Whitehead, *Adventures of Ideas* (New York: Free Press, 1993), part IV; see also Erin Manning, "Another Regard," in *The Lure of Whitehead*, ed. Nicholas Gaskill and A. J. Nocek (Minneapolis: University of Minnesota Press, 2014), 312–331; and Russell Duvernoy, *Affect and Attention After Deleuze and Whitehead: Ecological Attunement* (Edinburgh: Edinburgh University Press, 2021).
61. Hartmut Rosa, *Resonance: A Sociology of Our Relationship to the World* (Cambridge: Polity Press, 2019).
62. Haeckel, *Wonders of Life*, 33–34.
63. Whitehead, *Process and Reality*, 83–109. Though as I will argue later, we live in much more "democratic" societies than "monarchical." The difference Whitehead was trying to articulate might be one of degree more than anything else. As the new materialisms suggest, individual humans and humans in general are much less in control than we think.
64. On emergence theory, see Terrence Deacon, *Incomplete Nature: How Mind Emerged from Matter* (New York: Norton, 2012).

65. Catherine Keller, *Face of the Deep: A Theology of Becoming* (New York: Routledge, 2003).
66. Charles Taylor, *A Secular Age* (Cambridge, MA: Harvard University Press, 2007).
67. Bruno Latour, *The Politics of Nature: How to Bring the Sciences into Democracy* (Cambridge, MA: Harvard University Press, 2004), 187.
68. Though Gasman suggests Haeckel was a step along the way to Nazi ideology, I think (along with Robert Richards) that he has the history backward: That is, the Nazis used Haeckel in ways that to Gasman seem unproblematic, but in ways that Haeckel would have never allowed (were he even living during that time). Daniel Gasman, *The Scientific Origins of National Socialism* (New York: Routledge, 2017); Robert Richards, *The Tragic Sense of Life: Ernst Haeckel and the Struggle over Evolutionary Thought* Chicago: University of Chicago Press, 2009).
69. On the productive model, see Paul Thompson, *The Spirit of the Soil: Agriculture and Environmental Ethics* (New York: Routledge, 1995); see also Zimmerman, *Heidegger's Confrontation with Modernity*.
70. David Kaiser talks about this productive/reductive model and how those who sought wonder fought to break away from it; see David Kaiser, *How the Hippies Saved Physics: Science, Counterculture, and the Quantum Revival* (New York: Norton, 2011).
71. Val Plumwood makes this argument in *Environmental Culture: The Ecological Crisis of Reason* (New York: Routledge, 2002), 208–209.
72. J. Baird Callicott, *Thinking Like a Planet: The Land Ethic and the Earth Ethic* (New York: Oxford University Press, 2013).
73. Max Horkheimer and Theodor Adorno, *Dialectic of Enlightenment* (Stanford, CA: Stanford University Press, 2002).
74. Other perspectives such as those found in Tielhard de Chardin, process models (Alfred North Whitehead and Charles Hartshorne), and deep ecology (Arnae Naess) had, of course, been questioning such models all along.
75. Rachel Carson, *Silent Spring* (New York: Mariner, 1962).
76. Martin Heidegger, *The Question Concerning Technology, and Other Essays* (New York: Harper and Row, 1977).
77. This method of unlearning and unknowing is described in Whitney A. Bauman and Kevin O'Brien, *Environmental Ethics and Uncertainty: Wrestling with Wicked Problems* (New York: Routledge, 2019).

78. Lisa Sideris and Kathleen Dean Moore, *Rachel Carson: Legacy and Challenge* (Albany: SUNY Press, 2008).
79. Part of this argument regarding Carson was developed in Amanda Nichols and Whitney Bauman, "The Watery Depths of American Environmentalism: Marjory Stoneman Douglas, Rachel Carson, and Sylvia Earle," *Journal for the Study of Religion, Nature and Culture* 16, no. 3(2022): 394–425.
80. Diverting a bit from Mary-Jane Rubenstein's characterization of Haeckel's monism in *Pantheologies: Gods, Worlds. Monsters* (New York: Columbia University Press), 2018.
81. On the planetary context, see Whitney A. Bauman, *Religion and Ecology: Developing a Planetary Ethic* (New York: Columbia University Press, 2014); J. Baird Callicott, *Thinking Like a Planet: The Land Ethic and the Earth Ethic* (New York: Oxford University Press, 2013).
82. Chen, *Intoxicated*, 11.

I. EVOLUTION AND ECOLOGY: THE SCIENCES OF RADICAL IMMANENCE AND RELATIONALITY

1. Peter Harrison, *The Territories of Science and Religion* (Chicago: University of Chicago Press, 2015).
2. Walter Mignolo, *The Darker Side of Western Modernity: Global Futures, Decolonial Options* (Durham, NC: Duke University Press, 2011).
3. Thomas Berry, *The Great Work: Our Way Into the Future* (New York: Random House, 1999).
4. Timothy Morton, *Dark Ecology: For a Logic of Future Co-Existence* (New York: Columbia University Press, 2016).
5. The term *geohistory* is from Bruno Latour, *Facing Gaia: Eight Lectures on the New Climatic Regime* (Cambridge: Polity Press, 2017).
6. Brycchan Carey, "Clerical Naturalists in the Age of Enlightenment," www.brycchancarey.com, accessed September 10, 2025, www.brycchancarey.com/naturalists/index.htm.
7. In an 1888 letter from Darwin to Haeckel, Darwin warns Haeckel about making enemies of evolution due to Haeckel's strong rhetoric for evolution and against theology; see Franz M. Wuketits, "Mein Lieber Haeckel! Ernst Heckel, Charles Darwin und der Darwinismus," in *Darwin, Haeckel und die Folgen: Monismus in Vergangenheit*

und Gegenwart (Neustadt am Rübenberge, Germany: Angelika Lenz Verlag, 2006), 19–20. See also. on the formation of the Keplerbund against the Monistbund: Heiko Weber, *Monistische und antimonistische Weltanshauung: Eine Auswahlbibliographie* (Berlin, Germany: Verlag für Wissenschaft und Bildung, 2000), 14: "Parallel zu dem von Haeckel in Jena gegründeten *Phyletischen Museum* in Bad Godesberg ein *Museum für volkstümliche Naturkunde des Keplerbundes* gegründet wurde, die beide unter jeweils anderen weltanschaulichen Prämissen naturwissenschaftliche Volksbildung vermitteln wollten."

8. Carolyn Merchant, *Reinventing Eden: The Fate of Nature in Western Culture* (New York: Routledge, 2003), part I.
9. This is covered in greater detail in Whitney A. Bauman, *Theology Creation and Environmental Ethics: From* Creatio ex Nihilo *to* Terra Nullius (New York: Routledge, 2009).
10. Haeckel was clearly annoyed with Virchow on this point. Ernst Haeckel, *The Last Link: Our Present Knowledge of the Descent of Man* (London: Adam and Charles Black, 1898), 88: "Together with his [Virchow] colleagues Helmholtz the physicist, and Du Bois Reymond the physiologist, he has taken a leading place in the spreading of natural science; but, unfortunately, he did not take to the doctrine of Evolution, and for the last thirty years has been its declared antagonist, rarely missing an opportunity."
11. Jonathan Marks, "Why Were the First Anthropologists Creationists?" *Evolutionary Anthropology* 19 (2010): 222–226.
12. Haeckel wanted to avoid these ideologies imposed upon evolution, and argued against socialist or democratic understandings by showing how you could read just about any idea back into the evolutionary record. See, e.g., Erna Aescht, Gerhard Aubrecht, Erika Krauße, and Franz Speta, eds., *Welträtsel und Lebenswunder: Ernst Haeckel—Werk, Wirkung und Folgen* (Linz, Germany: OO Landesmuseums, 1998), 267: "so kann diese Tendenz nur eine aristokratische sein, durchaus keine demokratische, und am wenigsten eine sozialistische! Die Selektionstheorie lehrt, daß im Menschenleben wie im Tier- und Pflanzenleben überall und jederzeit nur eine kleine bevorzugte Minderzahl existieren und blühen kann; während die übergroße Mehrzahl darbt und mehr oder weniger frühzeitig elend zugrunde geht . . . Der grausame und schonungslose 'Kampf ums Dasein,' der überall in der

beendigen Natur wütet, und naturgemäß wüten muß, diese unaufhörliche und unerbittliche Konkurrenz alles Lebendigen, ist eine unleugbare Tatsache . . . Jedenfalls ist dieses Selektionsprinzip nichts weniger als demokratisch, sondern im Gegenteil aristokratisch im eigentlichsten Sinnes des Worts!"

13. See, e.g., Robert Niles Holt, "The Social and Political Ideas of the German Monist Movement, 1871–1914" (PhD Dissertation, Yale University, 1967).
14. Robert Richards, *The Tragic Sense of Life: Ernst Haeckel and the Struggle over Evolutionary Thought* (Chicago: University of Chicago Press, 2009), 235. Actually, Haeckel made the most pointed political use of his principle when he observed with amusement that those of royal blood would be chagrined to learn that during the first months of development, the human embryo of a noble or middle-class individual could not be distinguished from that of a dog. Haeckel also did so when he claimed that some races were akin to dogs and other animals, as seen in Erna Aescht, Gerhard Aubrecht, Erika Krauße, and Franz Speta, eds., *Welträtsel und Lebenswunder: Ernst Haeckel—Werk, Wirkung, und Folgen* (Linz, Germany: OO Landesmuseums, 1998), 233: "daß zwischen den höchst entwickelten Thierseelen' einerseits und den , tiefststehenden Menschenseelen' andererseits kein qualitativer, sondern nur ein quantitativer Unterschied existiere. . . . 'Sehr viele wilde Völker können nur bis zehn oder zwanzig zählen, während man einzelne sehr gescheute Hunde dazu gebracht hat, bis vierzig oder selbst über sechzig zu zählen.'"
15. Goethe, *Theory of Colors* (London: John Murray, 1840); Kate Rigby, *Topographies of the Sacred: The Poetics of Place in European Romanticism* (Charlottesville: University of Virginia Press, 2004), 20–22.
16. Indeed the "postmodern" condition only applies and makes sense for those who emerge out of a modern condition.
17. I'm reminded here of Mary Midgley's "many maps model"; Mary Midgley, *Science and Poetry* (New York: Routledge, 2002).
18. See, e.g., Heiko Weber and Olaf Breidbach, "Der Deutsche Monistbund: 1906 bis 1933," in *Darwin, Haeckel und die Folgen: Monismus in Vergangenheit und Gegenwart*, ed. A. Lenz and V. Mueller (Neustadt am Rübenberg, Germany: Angelika Lenz Verlag, 2006), 164: "Ziel der monistischen Ästhetik war demnach die Befreiung der Ästhetik

vom christlichen Dogma und ihre Begründung aus den Erkenntnissen der modernen Naturwissenschaften."

19. Ernst Haeckel, *Art Forms in Nature: The Prints of Ernst Haeckel* (Munich, Germany: Prestel-Verlag, 2015), 14.
20. Dalia Nassar, *Romantic Empiricism: Nature, Art and Ecology from Herder to Humboldt* (New York: Oxford University Press, 2022).
21. Gilles Deleuze and Felix Guattari, *A Thousand Plateaus: Capitalism and Schizophrenia* (Minneapolis: University of Minnesota Press, 1987).
22. Freya Matthews, *For the Love of Matter: A Contemporary Panpsychism* (Albany: State University of New York Press, 2003).
23. Max Horkheimer and Theodor Adorno, *Dialectic of Enlightenment* (Stanford, CA: Stanford University Press, 2002).
24. Eric Paul Jacobsen, *From Cosmology to Ecology: The Monist Worldview in Germany from 1770 to 1930* (Bern, Switzerland: Peter Lang, 2005), 208.
25. Londa Schiebinger, *Nature's Body: Gender in the Making of Modern Science* (Boston: Beacon, 1993), 11–39.
26. Sander Gilboff, *H.G. Bronn, Ernst Haeckel and the Origins of German Darwinism* (Cambridge, MA: MIT Press, 2008), 40.
27. Haraway, *Staying with the Trouble: Making Kin in the Chthulucene* (Durham, NC: Duke University Press, 2016), 58–98
28. Haeckel, *Kristallseelen: Studien über das anorganische Leben* (Leipzig, Germany: Alfred Kroner Verlag, 1917).
29. Olaf Breidbach, *Visions of Nature: The Art and Science of Ernst Haeckel* (Munich: Prestel, 2006), 277.
30. Gustav Theodor Fechner, *Nanna oder über das Seelenleben der pflanzen* (Germany: Edition Mabila, 2013), 55.
31. Terrance Deacon, *Incomplete Nature: How Mind Emerged from Matter* (New York: Norton, 2011).
32. Peter Wohlleben, *The Hidden Life of Trees: What They Feel, How they Communicate* (Vancouver: Greystone Books, 2016).
33. Nicholas Saul, "'Vorgänger Darwins' or 'Nachfolger Goethes'? Wilhelm Bölsche on Evolutionism in Goethe and Novalis or, Literature and the Two Cultures Then and Now," *Publications of the English Goethe Society* 84, no. 1 (2015): 76–91. See also: Bölche, *Love Life in Nature: The Story of the Evolution of Love*, 2 vols. (New York: Albert and Charles Boni, 1926).

34. Haeckel, *Monism as Connecting Science and Religion: The Confession of Faith of a Man of Science* (London: Adam and Charles Black, 1894), 15–16.
35. See also Advancing the Participatory Sciences, accessed January 15, 2026, https://participatorysciences.org/.
36. Ernst Haeckel, *Riddle of the Universe at the Close of the Nineteenth Century* (New York: Harper, 1900).
37. Marks, "Why Were the First Anthropologists Creationists?"
38. Breidbach, *Visions of Nature*, 137.
39. Stefano Harney and Fred Moten, *The Undercommons: Fugitive Planning and Black Study* (New York: Autonomedia, 2013).
40. Catherine Keller, *A Political Theology for the Earth* (New York: Columbia University Press, 2018), 64.
41. Donna Haraway makes this point about the false dichotomy between relativism and universalism. See, e.g., Donna Haraway, "Situated Knowledges: The Science Question in Feminism and the Privilege of Partial Perspective," *Feminist Studies* 14, no. 3 (Autumn 1988): 575–599.
42. Mignolo, *Darker Side of Western Modernity*, 505.

2. THEOLOGY, CRITICAL ROMANTICISM, AND THE CONSTRUCTION OF A NONREDUCTIVE NATURALISM

The epigraph is taken from Ernst Haeckel, *Gott-Natur (Theophysis), mit Kommenter Nachdruck von Olaf Breidbach und Uwe Hoßfeld* (Stuttgart, Germany: Franz Steiner Verlag, 2008), 69.

1. Andrea Wulf, *The Invention of Nature: Alexander von Humboldt's New World* (New York: Vintage 2015).
2. A source for what Bron Taylor identifies as "dark green religion."
3. See Kocku von Stuckrad, *A Cultural History of the Soul: Europe and North America from1870 to the Present* (New York: Columbia University Press, 2022); Kate Rigby, *Reclaiming Romanticism: Towards an Ecopoetics of Decolonization* (Oxford: Bloomsbury, 2021).
4. Gilles Deleuze and Felix Guattari, *A Thousand Plateaus: Capitalism and Schizophrenia* (Minneapolis: University of Minnesota Press, 1987).
5. Robert Richards excellent account of the development of the Romantic worldview oddly has little to say about Haeckel, which of course was made up for in the *Tragic Sense of Life: Ernst Haeckel and the*

Struggle Over Evolutionary Thought (Chicago: University of Chicago Press, 2009). His book on the emergence of Romantic understandings of nature provides way more detail than this one section ever could: *The Romantic Conception of Life: Science and Philosophy in the Age of Goethe* (Chicago: University of Chicago Press, 2002).

6. Ernst Haeckel, *The Riddle of the Universe at the Close of the Nineteenth Century* (New York: Harper, 1900), 193–194. See also Richards, *Tragic Sense of Life*, 350–351: "Just as Martin Luther, who 'with a mighty hand tore asunder the web of lies by the world-dominating papacy,' so in our day, Charles Darwin, with comparable overpowering might, has destroyed the ruling error doctrines of the mystical creation dogma and through his reform of developmental theory has elevated the whole sensibility, thought, and will of mankind onto a higher plane."
7. Haeckel, *Riddle of The Universe*, 18.
8. Robert Richards, "Ernst Haeckel and the Struggles over Evolution and Religion," *Annals of the History of Philosophy and Biology* 10: (2005): 91.
9. Heinrich Schmidt, ed., *Was Wir Ernst Haeckel Verdanken*, vol. 1 (Dresden, Germany: Fachbuchverlag, 2014), 21.
10. Hans Schwarz, "Darwinism Between Kant and Haeckel," *Journal of the American Academy of Religion* 43 (1980): 590–593.
11. Christoph Kockerbeck, ed., *Carl Vogt, Jacob Moleschott, Ludwig Büchner, Ernst Haeckel: Breifwechsel* (Marburg, Germany: Basilisken-Presse, 1999), 76.
12. Richards, *Tragic Sense of Life*, 327: "As against the likes of both Virchow and Bebel, one must, however, carefully consider Haeckel's ultimate conclusion about the positive political implications of evolutionary theory—namely, that there were none! Or at least, as he contended, the use of science to form policy for practical political decisions had to be hedged with a multitude of philosophical, social, and ethical considerations."
13. See Erik Lehnert, "Tiefes Gemüt, klarer Verstand und tapfere Kulturarbeit: Bruno Wille und der Freidrichshangener Dichterkreis als Ausgangspunkt monisticher Kulturpolitik im Kaiserreich," in *Darwin, Haeckel und die Folgen: Monismus in Vergangenheit und Gegenwart* (Neustadt am Rübenberge, Germany: Angelika Lenz Verlag, 2006), 261: "Vorbilder darin sind der 'Held' Giordano Bruno, der 'ethische Pantheist' Spinoza und v.a. der 'Dichterphilosoph' Goethe, dem Wille die Losung *Materie nie ohne Geist* verdankt."

14. Jeffrey Einboden, *Islam and Romanticism: Muslim Currents from Goethe to Emerson* (London: Oneworld Publications 2014).
15. Richards, *Tragic Sense of Life*, 352.
16. Richards, *Romantic Conception of Life;* see, e.g., 11, 223, 234, 506.
17. Mary-Jane Rubenstein, *Pantheologies: Gods, Worlds, Monsters* (New York: Columbia University Press, 2018).
18. Richards, *Tragic Sense of Life*, 126.
19. Ernst Haeckel, *The Wonders of Life: A Popular Study of Biological Philosophy* (New York: Harper, 1904), 82.
20. Wulf, *Invention of Nature.*
21. See, e.g., Ernst Haeckel, *Art Forms in Nature: The Prints of Ernst Haeckel* (Munich, Germany: Prestel-Verlag, 2015).
22. See Sander Gliboff's chapter, "What Kind of Racist Was Ernst Haeckel?" in Thomas Bach, Christina Brandt, and Florence Vienne, eds., *Exploring Ernst Haeckel's Work in the 21st Century* (University of Pittsburgh Press, forthcoming).
23. On "connective" versus "reductive" knowledge, see Freya Matthews, *For the Love of Matter: A Contemporary Panpsychism* (Albany: State University of New York Press, 2003).
24. Haeckel, *Riddle of the Universe*, 364.
25. See, e.g., Richards, *Tragic Sense of Life*, 34.
26. Wulf, *Invention of Nature*, 32.
27. Richards, *Romantic Conception of Life*, 12.
28. Von Stuckrad, *Cultural History of the Soul*, 10.
29. Richards, "Ernst Haeckel and the Struggles," 91.
30. James Miller, *China's Green Religion: Daoism and the Quest for a Sustainable Future* (New York: Columbia University Press, 2017).
31. Dalia Nassar, *Romantic Empiricism: Nature, Art and Ecology from Herder to Humboldt* (New York: Oxford University Press, 2022).
32. Wonder, Love, and Soul were all parts of Wilhelm Bölche and Gustav Fechner's understanding of a naturalistic worldview as well. See, e.g., Wilhelm Bölsche, *Von Wundern und Tieren: Neue Naturwissenschafliche Plaudereien* (Berlin: Deutsche Verlags-Anstalt, 1920); Wilhelm Bölsche, *Love Life in Nature: The Story of the Evolution of Love*, 2 vols (New York: Albert and Charles Boni, 1926); Gustav Fechner, *Nanna, Oder Uber das Sellenleben der Pflanzen* (Germany: Edition Mabila, 2013).
33. Wulf, *Invention of Nature*, 36.

34. Bölsche, *Love Life in Nature*, 725.
35. Nassar, *Romantic Empiricism*, 89.
36. Friedrich Wilhelm Schelling, *On the World Soul* (Hamburg, Germany: Perthes, 1798).
37. Fechner, *Nanna*.
38. Richards, *Romantic Conception of Life*, 11.
39. Andrea Mantell Seidel, *Isadora Duncan in the Twenty-First Century: Capturing the Art and Spirit of the Dancer's Legacy* (Jefferson, NC: McFarland, 2016), 65.
40. Donna Haraway, *Staying with the Trouble: Making Kin in the Chthulucene* (Durham, NC: Duke University Press, 2016), 49.
41. See, e.g., Zygmunt Bauman on Kant in *Liquid Love: On the Frailty of Human Bonds* (Cambridge: Polity Press, 2003), 125.
42. Kate Rigby, *Topography of the Sacred: The Poetics of Place in European Romanticism* (Charlottesville: University of Virginia Press), 38.
43. Haeckel, *Wonders of Life*, 69.
44. Schelling quoted in Rigby, *Topographies of the Sacred*, 45.
45. Ernst Haeckel, *Monism as Connecting Religion and Science: The Confession of Faith of a Man of Science* (London: Adam and Charles Black, 1894), 86.
46. Haeckel, *Riddle of the Universe*, 337.
47. See, e.g., Richards, *Tragic Sense of Life*, 45: "This kind of feeling of dependence and reverence in the presence of a greater power could rather easily be transferred onto a nature that displayed a commanding force and sublimity, which yet might yield up certain secrets to the microscope but which would trail off again into mysterious depths."
48. Richards, "Ernst Haeckel and the Struggles over Evolution," 89.
49. Haeckel, *Riddle of the Universe*, 194.
50. Haeckel, *Monism as Connecting Religion and Science*, 86.
51. E.g., see Haeckel, *Monism as Connecting Religion and Science*, 32, which argues, "We know that the innumerable varieties of animals and plants which during the course of millions of years have peopled our planet are all simply branches of one single genealogical tree; we know that the human race itself forms only one of the newest."
52. He writes, for example: "A general survey of the history of faith on the one hand and of science on the other clearly shows that the advance of the latter has always been accompanied by an increasing knowledge

of fixed natural laws and the shrinking superstition into an ever-lessening arena"; Haeckel, *Wonders of Life*, 56.

53. Rigby, *Topographies of the Sacred*, 49–50.
54. This is the type of multiculturalism that many postcolonial, decolonial, and critical race theories critique.
55. See Walter Mignolo, *The Dark Side of the Renaissance: Literacy, Territoriality, and Colonization* (Ann Arbor: University of Michigan Press, 2003); and Walter Mignolo, *The Darker Side of Western Modernity: Global Futures, Decolonial Options* (Durham, NC: Duke University Press, 2011).
56. Haeckel, *Monism as Connecting Religion and Science*, 15.
57. Haeckel, *Monism as Connecting Religion and Science*, 18–19.
58. Thomas Tweed, *Crossing and Dwelling: A Theory of Religion* (Cambridge, MA: Harvard University Press, 2008).
59. He argues, e.g., "Dogma would be replaced by a 'rationally based natural religion' whose 'highest commandment is love [and] a limitation of natural egoism for the betterment of human society.' Haeckel's monistic philosophy, braced by a natural religion of love, was meant to suggest the views of Goethe and Darwin; but his auditors would be encouraged by Virchow to hear only Feuerbach and Marx"; Richards, *Tragic Sense of Life*, 314–315).

3. ART FORMS OF NATURE: AESTHETICS "FROM BELOW"

A somewhat different version of this chapter appeared first as "Wonder and Ernst Haeckel's Aesthetics of Nature," in *Arts Religion and the Environment: Exploring Nature's Texture*, ed., Sigurd Bergmann and Forrest Clingerman (Leiden, the Netherlands: Brill, 2018), 61–83.

1. Ernst Haeckel, *Monism as Connecting Religion and Science: A Man of Science* (New York: Dossier Press, 1919) 18.
2. Marsha Morton, "From Monera to Man: Ernst Haeckel, Darwinismus, and Nineteenth-Century German Art," in *The Art of Evolution: Darwin, Darwinism, and Visual Culture*, ed. Barbara Larson and Fae Brauer (Hanover, NH: Dartmouth College Press, 2009), 65.
3. Ernst Haeckel, *Gott-Natur (Theophysis), mit Kommenter Nachdruck von Olaf Breidbach und Uwe Hoßfeld*, ed. Olaf Breidbach and Uwe Hoßfeld (Stuttgart, Germany: Franz Steiner Verlag, 2008), 18.

4. Haeckel, *Gott-Natur (Theophysis)*, 90.
5. Haeckel, 84.
6. Mary-Jane Rubenstein, *Strange Wonder: The Closing of Metaphysics and the Opening of Awe* (New York: Columbia University Press, 2010), 10–11.
7. Martin Heidegger, *Question Concerning Technology and Other Essays* (New York: Harper and Row, 1977), 3–35.
8. Olaf Breidbach, "Brief Instructions for Viewing Haeckel's Pictures," in *Art Forms in Nature: The Prints of Ernst Haeckel*, by Ernst Haeckel (Munich: Prestel-Verlag, 2015), 14.
9. Louise Economides, *The Ecology of Wonder in Romantic and Postmodern Literature* (New York: Palgrave Macmillan, 2016), 3.
10. Economides, *Ecology of Wonder*, 24.
11. As one author observes of Haeckel, "He assumed that the modifying effect of the environment on organisms is not random, but generally progressive, causing a lucky subset of each new generation of offspring to be slightly more perfect than the rest"; Eric Paul Jacobsen, *From Cosmology to Ecology: The Monist Worldview in Germany from 1770–1930* (Bern, Switzerland, Peter Lang, 2005), 98.
12. Heiko Weber, "Der Monismus als Theorie einer einheitlichen Weltanshauung am Beispiel der Positionen von Ernst Haeckel und August Forel," in *Monismus um 1900: Wissenschaftskultur und Weltanschauung*, ed. Paul Ziche (Berlin: Verlag für Wissenschaft und Bildung, 2000), 87.
13. Robert J. Richards describes his biogenetic law well: "A chief feature of Haeckel's evolutionary doctrine that supposedly distinguishes his views from those of Darwin is the principle of recapitulation. Haeckel put the principle thusly: The organic individual . . . repeats during the quick and short course of its individual development the most important of those changes in form that its ancestors had gone through during the slow and long course of their paleontological development according to the laws of inheritance and adaptation"; Robert J. Richards, *The Tragic Sense of Life: Ernst Haeckel and the Struggle over Evolutionary Thought* (Chicago: University of Chicago Press, 2008), 148.
14. Lorraine Daston and Katherine Park, *Wonders and the Order of Nature, 1150–1750* (New York: Zone Books, 1998), 109–134.
15. Homi K. Bhabha, *The Location of Culture* (New York: Routledge, 1984).
16. See, e.g., Michael Hogue, *The Tangled Bank: Toward an Ecotheological Ethics of Responsible Participation* (Eugene, OR: Wipf and Stock, 2008).

17. Bruno Latour, "Thou Shalt Not Freeze-Frame, or How Not to Misunderstand the Science and Religion Debate," in *Science, Religion and the Human Experience*, ed. James Proctor (New York: Oxford University Press, 2005), 27–48.
18. Donna Haraway, *Staying with the Trouble: Making Kin in the Chthulucene* (Durham, NC: Duke University Press, 2016).
19. Bruno Latour, *After Lockdown: A Metamorphosis* (Cabridge: Polity Press, 2021), 53.
20. Magnus Hirschfeld, "Ernst Haeckel und die Sexualwissenschaft" in Heinrich Schmidt, ed. *Was wir Ernst Haeckel verdanken: Ein Buch der Verehrung und Dankbarkeit, 2. Band* (Leipzig: Verlag Unesma, 1914), 282–284. There are several letters from Hirschfeld to Haeckel in which Hirschfeld thanks him for his contributions to his own work. Speaking of his insights into anthropology, for instance, Hirschfeld writes, "Ich erblicke in ihnen wertvolle Hinweise für weiteres wissenschafliches Forschen auf diesem nach so zahlreichen Richtungen und Beziehungen hin zu ergründenen Wissensgebiet, dessen Erschliessung im Geists des Ihr wissenschaftliches Schaffen charakterisierenden Strebens, empirische Forschungsergebnisse den grossen Zusammenhängen des Naturgeschehens einzureihen, meine Lebensarbeit, soweit meine Kräfte ausreichen, gewidmet sein soll" (letter from Hirschfeld to Haeckel, May 6, 1912; in the Ernst Haeckel Haus, Jena, Germany). Hirschfeld basically tells Haeckel here that his work has given him many pointers and that he is inspired by Haeckel to do the best scientific work he can do (here in the area of sexuality, more specifically).
21. Breidbach, "Brief Instructions," 14.
22. Translation mine. "Die Symmetrie der Formen, die einfach Passung mit seinem ästhetischen Empfinden sichert für Haeckel eine Empfindungseinheit des Naturalen, die er als Beseeltheit alles Existenten interpretiert"; Haeckel, *Gott-Natur (Theophysis)*, 33.
23. See, e.g., Catherine Keller, *Cloud of the Impossible: Negative Theology and Planetary Entanglement* (New York: Columbia University Press, 2014).
24. See Kate Rigby on the history of F. Schlegel's use of the term, and her own retrieval of it in *Reclaiming Romanticism: Towards an Ecopoetics of Decolonization* (Oxford: Bloomsbury, 2021).
25. Haraway, *Staying with the Trouble*, 33.

26. See F. W. J. Schelling, *Bruno: On the Natural and Divine Principle of Things*, trans. Michael Vater (Albany: State University of New York Press, 2000).
27. Jack Halberstam, *The Queer Art of Failure* (Durham, NC: Duke University Press, 2011).
28. See, e.g., "No science of any kind whatever consists solely in the description of observed facts"; Ernst Haeckel, *The Wonders of Life: A Popular Study of Biological Philosophy* (New York: Harper, 1905), 5–6.
29. On Kant's false move to the a priori/ideal in his epistemology, Haeckel writes, "This dogma is erroneously built on the correct idea that our knowledge, obtained through the senses, is imperfect; it extends only so far as the specific energy of the senses and the structure of the phronema admit. But it by no means follows that it is a mere illusion, and least of all that the external world exists only in our ideas"; Haeckel, *Wonders of Life*, 69.
30. Morton, "From Monera to Man," 77–82.
31. Morton, 63.
32. Ernst Haeckel, *History of Creation: Or the Development of the Earth and Its Inhabitants by the Action of Natural Causes*, vol. 1 (New York: D. Appleton & Company, 1876), chap. 15.
33. Ernst Haeckel, *Kristallseelen: Studien über das anorganische Leben* (Leipzig, Germany: Alfred Kroner Verlag, 1917).
34. Haeckel, *Wonders of Life*, 33–34.
35. James Miller, *China's Green Religion: Daoism and the Quest for a Sustainable Future* (New York: Columbia University Press, 2017).
36. For a brief description of emergence, see Ursula Goodenough and Terrence Deacon, "The Sacred Emergence of Nature," in *The Oxford Handbook of Science and Religion*, ed. Phillip Clayton (New York: Oxford University Press, 2008), 853–871.
37. See, e.g., Kocku von Stuckrad, *A Cultural History of the Soul: Europe and North America: From 1850 the Present* (New York: Columbia University Press, 2022), 96–105.
38. Haeckel's support for arts and education was in part due to the fact that evolution could be regressive or progressive; see, e.g., Erika Ellis, "The Reception of Ernst Haeckel in America: Monism and the Prophet of Evolution" (undergraduate thesis, University of Washington, Seattle, 1998), 70. On the other hand, his understanding of evolution does seem

to imply a progressive tone toward more complexity and advancement; see e.g., Eric Paul Jacobsen, *From Cosmology to Ecology: The Monist Worldview in Germany from 1770 to 1930* (Bern: Peter Lang, 2005), 98.

Sander Gliboff, *H. G. Bron, Ernst Haeckel and the Origins of German Darwinism: A Study in Translation and Transformation* (Cambridge, MA: MIT Press, 2008), 155: "Haeckel's new uses of 'Entwicklung' referred to an open-ended and unpredictable process, not a chain of fully predetermined forms. When he equated traditional types with Darwinian ancestors or lineages, he was not misunderstanding or undermining the Darwinian conception, but indicating to morphologists and systematists where and how to apply it. Phenomena such as unity of type were explained in new ways, even if they were not given new names. Similarly, when Haeckel spoke of progress and perfection, he referred to historical trends like those Bronn had described and analyzed, not to an idealized sequence of forms, and he explained them in terms of adaptation and selection."

39. Hogue, *Tangled Bank.*
40. Kate Rigby, *Topographies of the Sacred: The Poetics of Place in European Romanticism* (Charlottesville: University of Virginia Press), 48–49.
41. Gilles Deleuze and Felix Guattari, *A Thousand Plateaus: Capitalism and Schizophrenia* (Minneapolis: University of Minnesota Press, 1987), 93–98.
42. Alfred North Whitehead, *Science and the Modern World* (New York: MacMillan, 1925), 51–58.
43. See, e.g., Jane Bennett, *Vibrant Matter: A Political Ecology of Things* (Durham, NC: Duke University Press, 2010); Karen Barad, *Meeting the Universe Halfway: Quantum Physics and the Entanglement of Matter and Meaning* (Durham, NC: Duke University Press, 2007).
44. Haeckel, *Gott-Natur (Theophysis)*, 22.
45. Bruno Latour, *If We Lose the Earth, We Lose Our Souls* (Cambridge: Polity Press, 2024).
46. Morton, "Monera to Man," 64.
47. On emergent teleology, see Terrence Deacon, *Incomplete Nature: How Mind Emerged from Matter* (New York: Norton, 2012).
48. Russel Duvernoy's idea of ecological attunement in conversation with Deleuze and Whitehead (among others) suggests this type of

ontology. See his *Affect and Attention After Deleuze and Whitehead: Ecological Attunement* (Edinburgh: Edinburgh University Press, 2021).

49. Baird Callicott, *Thinking Like a Planet: The Land Ethic and the Earth Ethic* (New York: Oxford University Press, 2014), 225–226.

4. THE UGLY SIDE OF COMMON GROUNDS

The epigraph of this chapter is taken from Karl Rosenkranz, *Aesthetics of the Ugly* (London: Bloomsbury, 2015), 103.

1. Walter Mignolo, *The Darker Side of Western Modernity: Global Futures, Decolonial Options* (Durham, NC: Duke University Press, 2011).
2. Rosenkranz, *Aesthetics of the Ugly* (London: Bloomsbury, 2015), 32.
3. In a letter to Frida von Uslar-Gleichen, Haeckel laments that if there were a good God, he would not let such horrible things happen. See Norber Elsner, *Das Ungelöste Welträtsel: Frida von Uslar-Gleichen und Ernst Haeckel Briefe und Tagebücher 1898–1903, Band 3* (Göttingen, Germany: Wallstein Verlag, 1996), 416: "Wenn es aber wirklich einen 'guten lieben Gott' giebt, so würde dieser Gott des Wahren, Guten und Schönen segnend seine Arme über uns breiten, unsere beiden Hände in einander legen, und sagen: 'Ihr lieben beiden seltsamen Kinder, Ich segne u. behüte Euch!' Und unsere herrliche Mutter Natur würde dazu frohlockend in ihre gütigen Hände klatschen, und rufen: 'Das ist herrlich! Ihr beiden außerordentlichen Lieblinge, die ich so reich mit besonderer Liebe ausgestattet, habt Euch endlich doch gefunden, und nun bleibt glückselig beisammen!'" See also Robert Richards, *Tragic Sense of Life: Ernst Haeckel and the Struggle over Evolutionary Thought* (Chicago: University of Chicago Press, 2009), 167: "Haeckel felt the urgency of its realization not only as balm to a terrible loss but also to lay out a new, reliable path through the tangle of superstition, both religious and scientific, that insinuated false hope and choked off the voice of reason. The path he envisioned would terminate in a new, modern age, one in which religion would be transformed into Darwinian understanding and the individual would find rest in the bosom of Goethean nature."
4. You can even see his theodical questions beginning with the illness and death of his aunt Bertha. While still at university, Haeckel writes to his parents: "Sehr viel haben wir an unsre arme, arme Tante B*ertha*

gedacht, u*nd* ihre Leiden von Herzen bedauert. Aber wenn ihr das nur etwas hülfe! Es ist doch gar zu hart u*nd* grausames Leid, nachdem es ihr im Sommer so viel besser gegangen, nun wieder ganz elend u*nd* flach dazuliegen. Da möchte man wirklich oft den lieben Gott fragen, wie er so etwas zugeben kann!"; Ernst Haeckel, *Ausgewählte Briefwechsel. Historisch-kritische Ausgabe. Band 1: Familien Korrespondenz Februar 1839–April 1845*, ed. Roman Göbel, Gerhard Müller, Claudia Taszus, Thomas Bach, Jens Pahnke and Kathrin Polenz (Stuttgart, Germany: Franz Steiner, 2017), 366.

5. Todd H. Weir, ed., *Monism: Science, Philosophy, Religion and the History of a Worldview* (New York: Palgrave, 2012), 6.
6. See, Erika Ellis, "The Reception of Ernst Haeckel in America: Monism and The Prophet of Evolution." Undergraduate thesis, University of Washington, Seattle, 1998, 70: "Haeckel theorized, as did many other freethinkers, that the future evolution of the human race was not ensured and instead depended upon its further enlightenment and humanitarianism. Haeckel demanded an active philosophy because evolution could be regressive and not progressive in certain unfortunate circumstances."
7. Jürgen Sandmann, *Der Bruch mit der Humanitären Tradition: Die Biologisierung der Ethik bei Ernst Haeckel und anderen Darwinisten seiner Zeit* (Stuttgart, Germany: Gustav Fischer Verlag, 1990), 109–113.
8. Mary-Jane Rubenstein, *Astrotopia: The Dangerous Religion of the Corporate Space Race* (Chicago,: University of Chicago Press, 2022), 5.
9. Bruno Latour, *If We Lose the Earth, We Lose Our Souls* (Cambridge: Polity Press, 2024).
10. Michel Foucault, *Discipline and Punish: The Birth of the Prison* (New York: Random House, 1977).
11. Carl Schmitt, *The Concept of the Political* (Chicago: University of Chicago Press, 1996). See, e.g., Clayton Crockett and Jeffrey Robbins, *Religion, Politics, and the Earth: The New Materialism* (New York: Palgrave, 2012).
12. The postcolonial critiques of science are multiple. See, e.g., Sandra Harding, ed., *The Postcolonial Science and Technology Studies Reader* (Durham, NC: Duke University Press, 2011). Environmental historians have also argued similarly. See, e.g., Carolyn Merchant, *Reinventing Eden: The Fate of Nature in Western Culture* (New York:

Routledge, 2003). All manner of critical race theories, Indigenous studies, and decolonial studies have also discussed the colonizing metaphors of science. See, e.g., Michelle M. Wright, *Physics of Blackness: Beyond the Middle Passage Epistemology* (Minneapolis: University of Minnesota Press, 2015); Laurelyn Whitt, *Science, Colonialism and Indigenous Peoples: The Cultural Politics of Law and Knowledge* (New York: Cambridge University Press, 2009); Mignolo, *Darker Side of Western Modernity*.

13. Sandra Harding, *Objectivity and Diversity: Another Logic of Scientific Research* (Chicago: University of Chicago Press, 2015).
14. Harding, *Objectivity and Diversity*.
15. Martin Heidegger, *The Question Concerning Technology and Other Essays* (New York: Harper and Row, 1977).
16. Eduardo Kohn, *How Forests Think: Toward an Anthropology Beyond the Human* (Berkeley: University of California Press, 2013), 183.
17. Hannah Arendt, *The Life of the Mind* (New York: Harcourt, 1978).
18. Terrance Deacon, *Incomplete Nature: How Mind Emerged from Matter* (New York: Norton, 2012).
19. Gilles Deleuze and Felix Guattari, *A Thousand Plateaus: Capitalism and Schizophrenia* (Minneapolis: University of Minnesota Press, 1987).
20. Donna Haraway, *Simians, Cyborgs and Women: The Reinvention of Nature* (New York: Routledge, 1991).
21. Daniel Gasman, *The Scientific Origins of National Socialism* (New York: Routledge, 2004).
22. Stephen J. Gould, *Ontogeny and Phylogeny* (Cambridge, MA: Harvard University Press, 1977); Richards, *Tragic Sense of Life*.
23. A brief look, for instance, of the earliest forms of comparative religious studies in the nineteenth century shows how early anthropologists and sociologists understood religion as on a scale from "primitive" to "monotheistic," with monotheism being the most evolved and at times superseded by the emerging modern scientific paradigm.
24. Benedict Anderson, *Imagined Communities: Reflections on the Origin and Spread of Nationalism* (New York: Verso, 1983).
25. On the idea of "common grounds" versus "foundations," see Catherine Keller and Laurel Kearns, "Introduction: Grounding Theory-Earth in Religion and Philosophy," in *EcoSpirit: Religions and*

Philosophies for the Earth, ed. Laurel Kearns and Catherine Keller (New York: Fordham University Press, 2007), 1–20.

26. Loretta J. Ross, *Calling In: How to Start Making Change with Those You'd Rather Cancel* (New York: Simon and Schuster, 2025).
27. Richards, *Tragic Sense of Life*, 447.
28. Richards, 451.
29. Haeckel was on the side of the nineteenth-century scientists who sought a more holistic form of science. See, e.g., Rudolf Bährmann, "Haeckels ökologie-Begriff: Inhalt, Deutung, Bedeutung," in Arnher E. Lenz und Volker Mueller, eds., *Darwin, Haeckel und die Folgen: Monismus in Vergangenheit und Gegenwart* (Neustadt am Rübenberge, Germany: Angelika Lenz Verlag, 2006), 99: "Zwei Hauptrichtungen werden sehr bald sichtbar, 1. Eine reduktionistische, die auf analytische Weise Erkenntnisse zu erlangen trachtet, und 2. Eine holistische, die sich bemüht, durch Synthese von Einzelbefunden auf eine höhere Ebene der Erkenntnis, einer Synopsis von Formen- und Funktionszusammenhängen vorzudringen."
30. Ernst Haeckel, *Wonders of Life*, 69.
31. Todd H. Weir, ed., *Monism: Science, Philosophy, Religion and the History of a Worldview* (New York: Palgrave, 2012), 3.
32. Haeckel, *Riddle of the Universe*, 9: "Never will our government improve until it casts off the fetters of the Church and raises the views of the citizens on man and the world to a higher level by a general scientific education. That does not raise the question of any special form of constitution. Whether a monarchy or a republic be preferable, whether the constitution should be aristocratic or democratic, are subordinate questions in comparison with the supreme question: Shall the modern civilized state be spiritual or secular? Shall it be theocratic —ruled by the irrational formulae of faith and by clerical despotism—or nomocratic —under the sovereignty of rational laws and civic right."
33. Michael Zimmerman, *Heidegger's Confrontation with Modernity: Technology, Politics, and Art* (Bloomington: Indiana University Press, 1990).
34. Val Plumwood, *Environmental Culture: The Ecological Crisis of Reason* (New York: Routledge, 2002), 209: "The extermination programs were in many respects an extreme *expression* of modernity and its rational capacity to scapegoat, marginalize, and eliminate; both the programs and Nazi thinking in general involved strong modernist

elements of human/animal dualism, instrumental/bureaucratic order and hyper-rationalism."

35. Heidegger, *Question Concerning Technology*.
36. Zimmerman, *Heidegger's Confrontation with Modernity*, 189.
37. Vandana Shiva, *Staying Alive: Women, Ecology and Development* (London: Zed Books 1988); Rosemary Radford Ruether, *Integrating Ecofeminism, Globalization, and World Religions* (Landham, MD: Roman and Littlefield, 2005).
38. Rachel Carson, *Silent Spring* (New York: Houghton Mifflin, 1962).
39. Robert Richards, "The German Reception of Darwin's Theory, 1860–1945," in *The Cambridge Encyclopedia of Darwin and Evolutionary Thought* (Cambridge: Cambridge University Press, 2013), 5.
40. Already in 1879 you could see these culture wars forming. In a letter to the Prussian Kulturminister, Wilhelm Freiherr von Hammerstein (former editor of the *Kreuzzeitung*) wrote against the teaching of Haeckel and Darwin in the schools: "wenn in unserem Vaterland eine Generation heranwächst, deren Glaubensbekenntnis der Atheismus und der Nihilismus, deren politische Anschauung der Kommunismus ist"; see Erika Krauße, *Ernst Haeckel: Biographien Hervorragender Naturwissenschaftler, Techniker und Mediziner* (Leipzig, Germany: Teubner Verlagsgesellschaft, 1987), 255.
41. See, e.g., Baird Callicott, *Thinking Like a Planet: The Land Ethic and the Earth Ethic* (New York: Oxford University Press, 2013).
42. Bruno Latour, *Facing Gaia: Eight Lectures on the New Climatic Regime* (Cambridge: Polity Press, 2017).
43. Donovan Schaefer, *Wild Experiment: Feeling Science and Secularism After Darwin* (Durham, NC: Duke University Press, 2022), 5.
44. Horst W. J. Rittel and Melvin M. Webber, "Dilemmas in a General Theory of Planning," *Policy Sciences* 4, no. 2 (June 1973): 155–169.
45. Ernst Haeckel's *Riddle of the Universe* was confident about solving those riddles.
46. What Mayra Rivera so beautifully describes as a "poetics of the flesh"; Mayra Rivera, *Poetics of the Flesh* (Durham, NC: Duke University Press, 2015).
47. This is a play on Andreas Weber's "erotic ecology," see his: *Matter and Desire: An Erotic Ecology* (White River Junction, VT: Chelsea Green, 2017). I talk about some of this in my article, "Critical Planetary

Romanticism: Ecology, Evolution and Erotic Thinking," *Journal for the Study of Religion, Nature and Culture* 15, no. 1 (2021): 33–52.

48. See, e.g., Karen Barad, *Meeting the Universe Halfway: Quantum Physics and the Entanglement of Matter and Meaning* (Durham, NC: Duke University Press, 2007); William Connolly, *Climate Machines, Fascist Drives, and Truth* (Durham, NC: Duke University Press, 2019); Jane Bennett, *Vibrant Matter: A Political Ecology of Things* (Durham, NC: Duke University Press, 2010); Deleuze and Guattari, *Thousand Plateaus*.
49. Freya Matthews, *For the Love of Matter: A Contemporary Panpsychism* (Albany: State University of New York Press, 2003); Val Plumwood, *Environmental Culture*; Karen Bray, Heather Eaton and Whitney Bauman, *Earthly Things: Immanence, New Materialisms, and Planetary Thinking* (New York: Fordham University Press, 2023).
50. For Haeckel, his refusal of theism and dualism led to an embrace of a monism, which is really just another container turning mono-theism into mono-naturalism. In other words, his trap was in thinking a reversal without seriously undoing the edges of knowledge claims and embracing uncertainty. This, thought Haeckel the scientist but maybe not Haeckel the philosopher or artist, would provide us with objective knowledge of the world.
51. This is the point of affect theories such as found in Donovan Schaefer, *Religious Affects: Animality, Evolution, and Power* (Durham, NC: Duke University Press, 2015).
52. Bruno Latour, "Thou Shalt Not Freeze-Frame, or How Not to Misunderstand the Science and Religion Debate," in *Science, Religion and the Human Experience*, ed. James Proctor (New York: Oxford University Press, 2005), 99–123.
53. These spaces are like black holes of our mental landscapes sucking in earthly bodies and destroying them in the process.
54. See, e.g., Catherine Keller, *Cloud of the Impossible: Negative Theology and Planetary Entanglement*. New York: Columbia University Press, 2014. Ursula Goodenough, *The Sacred Depths of Nature: How Life Has Emerged and Evolved*, 2nd rev. ed. (New York: Oxford University Press, 2023).
55. Jack Halberstam, *The Queer Art of Failure* (Durham, NC: Duke University Press, 2011).

56. See, e.g., Aph Ko, *Racism as Zoological Witchcraft: A Guide to Getting Out* (Brooklyn, NY: Lantern Books, 2019).
57. See Karen Bray, *Grave Attending: A Political Theology for the Unredeemed* (New York: Fordham University Press, 2020). She argues for "staying with the unredeemed," and the unhappy, depressive spots, in order to attend to (and reattune to) abjected bodies.
58. Marcella Althaus-Reid, *Indecent Theology: Theological Perversions in Sex, Gender and Politics* (New York: Routledge, 2007).

5. REATTUNEMENTS TO THE PLANETARY COMMUNITY

1. Including myself in Whitney Bauman, *Religion and Ecology: Developing a Planetary Ethic* (New York: Columbia University Press, 2014).
2. Mary Evelyn Tucker, *Worldly Wonder: Religions Enter Their Ecological Phase* (Peru, IL: Open Court, 2003), 9.
3. Bruno Latour, *Facing Gaia: Eight Lectures on the New Climatic Regime*, trans. Catherine Porter (Cambridge: Polity Press, 2017), 73.
4. Stefano Harney and Fred Moten, *The Undercommons: Fugitive Planning and Black Study* (New York: Autonomedia, 2013).
5. In several letters Haeckel approvingly remarks on Alexander Tille's *Von Darwin bis Nietzsche: Ein Buch Entwicklungsethik*, for instance. See "Alexander Tille an Ernst Haeckel, Glasgow, 6. January 1895," Ernst Haeckel Online Briefedition, accessed January 17, 2026, https://haeckel-briefwechsel-projekt.uni-jena.de/en/document.xhtml?id=b_18273.
6. Friedrich Nietzsche, *The Gay Science*, trans. Walter Kaufmann (New York: Vintage Books, 1974).
7. Friedrich Nietzsche, *Thus Spake Zarathustra*, trans. Thomas Common (New York: Macmillan, 1911).
8. Jane Bennet, *Vibrant Matter: A Political Ecology of Things* (Durham, NC: Duke University Press, 2010).
9. Whitney A. Bauman, *Theology, Creation and Environmental Ethics: From* Creatio ex Nihilo *to* Terra Nullius (New York: Routledge, 2009).
10. Mary-Jane Rubenstein, *Pantheologies: Gods, Worlds, Monsters* (New York: Columbia University Press, 2018), xx.
11. Catherine Keller, *God and Power: Counter-Apocalyptic Journeys* (Minneapolis, MN: Fortress Press, 2005).

12. Michel Foucault, *Power/Knowledge: Selected Interviews and Other Writings, 1972–1977* (New York: Random House, 1988).
13. Hannah Arendt, *The Human Condition* (Chicago: University of Chicago Press, 1958).
14. Sylvia Wynter, *On Being Human as Praxis* (Durham, NC: Duke University Press, 2015).
15. Aph Ko, Racism as Zoological Witchcraft: A Guide to Getting Out (Brooklyn, NY: Lantern Books, 2019).
16. Carol Wayne White, *Black Lives and Sacred Humanity: Toward an African American Religious Naturalism* (New York: Fordham University Press, 2016).
17. Margaret Kovach, *Indigenous Methodologies: Characteristics, Conversations, and Contexts*, 2nd ed. (Toronto: University of Toronto Press, 2021).
18. Donna Haraway, *Staying with the Trouble: Making Kin in the Chthulucene* (Durham, NC: Duke University Press, 2016).
19. James Miller makes this point in his book *China's Green Religion: Daoism and the Quest for a Sustainable Future* (New York: Columbia University Press, 2017).
20. Ursula Goodenough, *The Sacred Depths of Nature: How Life Has Emerged and Evolved*, 2nd rev. ed. (New York: Oxford University Press, 2023), 102.
21. See, e.g., Haeckel, *Kristallseelen: Studien über das anorganische Leben* (Leipzig, Germany: Alfred Kroner Verlag, 1917) and *The Riddle of the Universe at the Close of the Nineteenth Century* (New York: Harper, 1900).
22. Eric Paul Jacobsen, *From Cosmology to Ecology: The Monist Worldview in Germany from 1770 to 1930* (Bern, Switzerland: Peter Lang Press, 2005), 102.
23. On emergence theory, see Terrence Deacon, *Incomplete Nature: How Mind Emerged from Matter* (New York: Norton, 2012).
24. Ivone Gebara, *Longing for Running Water: Ecofeminism and Liberation* (Minneapolis, MN: Fortress Press, 1999).
25. Karen Barad, *Meeting the Universe Halfway: Quantum Physics and the Entanglement of Matter and Meaning* (Durham, NC: Duke University Press, 2007); Judith Butler, *Gender Trouble: Feminism and the Subversion of Identity* (New York: Routledge, 1990).
26. Hannah Arendt, *Sokrates. Apologie der Pluralität* (Berlin: Matthes and Seitz, 2016).

27. Goodenough, *Sacred Depths of Nature*, 31–36.
28. Also emphatically argued by Friedrich Schleiermacher in *On Religion: Talks to its Cultured Despisers* (1799), Christian Classics Ethereal Library, accessed January 15, 2026, www.ccel.org/ccel/s/schleiermach/religion/cache/religion.pdf.
29. Bauman, *Theology, Creation and Environmental Ethics*; and *Religion and Ecology*.
30. Bruno Latour, *Down To Earth: Politics in the New Climactic Regime* (Cambridge: Polity Press, 2018).
31. Benedict Anderson, *Imagined Communities: Reflections on the Origin and Spread of Nationalism* (New York: Verso, 1983).
32. Whitney Bauman, "Developing a Critical Romanticism" (blog), *Counterpoint: Navigating Knowledge*, August 14, 2019, www.counterpointknowledge.org/developing-a-critical-romanticism/. See also Whitney Bauman, "Developing a Critical Planetary Romanticism: Re-Attuning to the Earth," in *Religion, Materialism, and Ecology*, ed. Sigurd Bergmann, Kate Rigby, and Peter Manley Scott (New York: Routledge, 2023), 13–28.
33. There is a growing body of literature on "religion and infrastructures." See. E.g., Benjamin Kirby, "Religious Infrastructure: Designations, Transformations, Entanglements," *Religion, State and Society* 52, nos. 2–3 (2024): 96–113.
34. Rubenstein, *Astrotopia*.
35. Jan Millsapps, "The Arduous Journey of Staying Home" (blog), *Counterpoint, Navigating Knowledge*, March 26, 2025, www.counterpointknowledge.org/the-arduous-journey-of-staying-home/.
36. Sarah McFarland Taylor, "'F*ck Earth': Unmasking Mars Colonization Marketing, from Planetary Perceived Obsolescence to Apocalyptic 'New Earth' Rhetoric," *Journal of Religion, Media and Culture* 11 (2022): 54–84.
37. Douglas Rushkoff, *Survival of the Richest: Escape Fantasies of the Tech Billionaires* (London: Scribe Publications, 2022).
38. Taylor, "F*ck Earth."
39. See, e.g., "No science of any kind whatever consists solely in the description of observed facts," in Ernst Haeckel, *The Wonders of Life: A Popular Study of Biological Philosophy* (New York: Harper, 1905), 5–6.

40. Bruno Latour, *The Politics of Nature: How to Bring Sciences into Democracy* (Cambridge, MA: Harvard University Press, 2004).
41. Wendy Wheeler, *Expecting the Earth: Life, Culture, Biosemiotics* (London: Lawrence and Wishart, 2016).
42. Jack Halberstam, *Wild Things: The Disorder of Desire* (Durham, NC: Duke University Press, 2020), 18.
43. Donna Haraway, *Simians, Cyborgs, and Women: The Reinvention of Nature* (New York: Routledge 1991).
44. On Kant's false move to the a priori/ideal in his epistemology, Haeckel writes, "This dogma is erroneously built on the correct idea that our knowledge, obtained through the senses, is imperfect; it extends only so far as the specific energy of the senses and the structure of the phronema admit. But it by no means follows that it is a mere illusion, and least of all that the external world exists only in our ideas"; Haeckel, *Wonders of Life*, 69.
45. Walter Mignolo, *The Darker Side of Western Modernity: Global Futures, Decolonial Options* (Durham, NC: Duke University Press, 2011), 273.
46. Jack Halberstam, *The Queer Art of Failure* (Durham, NC: Duke University Press, 2011).
47. Haeckel's support for arts and education was in part due to the fact that evolution could be regressive or progressive; see, e.g., Erika Ellis, "The Reception of Ernst Haeckel in America: Monism and The Prophet of Evolution" (undergraduate thesis, University of Washington, Seattle, 1998), 70. On the other hand, his understanding of evolution does seem to imply a progressive tone toward more complexity and advancement; see e.g., Eric Paul Jacobsen, *From Cosmology to Ecology: The Monist Worldview in Germany from 1770 to 1930* (Bern, Switzerland: Peter Lang, 2005), 98.
48. Mignolo, *Darker Side of Modernity*, 217.
49. Michelle Bastian, "Encountering Leatherbacks in Multispecies Knots of Time" in *Extinction Studies: Stories of Time, Death and Generations*, ed. T. van Dooren, D. Bird Rose, and M. Chrulew (New York: Columbia University Press, 2017), 149–186.
50. Gilles Deleuze and Felix Guattari, *A Thousand Plateaus: Capitalism and Schizophrenia* (Minneapolis: University of Minnesota Press, 1987).

51. Isabelle Stengers, "The Cosmopolitical Proposal," in *Making Things Public*, edited by Bruno Latour and Peter Wiebel (Cambridge, MA: MIT Press, 2005).

6. THE EMERGENCE OF A CRITICAL PLANETARY ROMANTICISM FOR EARTH'S UNCERTAIN FUTURE

A modified version of the section "Critical Planetary Romanticism" in this chapter appears in Whitney Bauman, "Developing A Critical Planetary Romanticism: Re-attuning to the Earth," in *Religion, Materialism, and Ecology*, ed. Sigurd Bergmann, Kate Rigby, and Peter Manley Scott (New York: Routledge, 2023), 21–24. Some of the ideas in the "Carson's Water Ethic and Critical Planetary Thinking for Planetary Politics" section were developed in: Amanda Nichols and Whitney Bauman, "The Watery Depths of American Environmentalism: Marjory Stoneman Douglas, Rachel Carson, and Sylvia Earle," *Journal for the Study of Religion, Nature and Culture* 16, no. 3 (2022): 294–425; and Whitney Bauman and Kevin O'Brien, *Environmental Ethics and Uncertainty: Wrestling with Wicked Problems* (New York: Routledge, 2019), chap. 2.

1. Though I find Daniel Gasman's book a bit unfair to Haeckel on this, I do find his indictment of science in general to be relevant: *The Scientific Origins of National Socialism* (New York: Routledge, 2017).
2. Lisa Sideris offers a good critique of this; see *Consecrating Science: Wonder, Knowledge and the Natural World* (Berkeley: University of California Press, 2017). On the "scientification" of knowledge, see Kocku von Stuckrad, *The Scientification of Religion: An Historical Study of Discursive Change, 1800–2000* (Berlin: De Gruyter, 2014).
3. Donna Haraway, *Staying with the Trouble: Making Kin in the Chthulucene* (Durham, NC: Duke University Press, 2016).
4. Catherine Keller and Laurel Schneider, eds., *Polydoxy: Theology of Multiplicity and Relation* (New York: Routledge, 2011).
5. Jack Halberstam, *The Queer Art of Failure* (Durham, NC: Duke University Press, 2011).
6. See Whitney Bauman, "Queering Authority in Science and Religion," in *Unsettling Science and Religion: Contributions and Questions from Queer Studies*, ed. Lisa Stenmark and Whitney Bauman (Lanham, MD: Lexington Books, 2018), 89–106.

7. Michael E. Zimmerman, *Heidegger's Confrontation with Modernity: Technology, Politics and Art* (Bloomington: Indiana University Press, 1990).
8. Bruno Latour, *Politics of Nature: How to Bring the Sciences into Democracy* (Cambridge, MA: Harvard University Press, 2004).
9. Jacques Derrida, *The Politics of Friendship* (New York: Verso, 1994).
10. Catherine Keller, *God and Power: Counterapocalyptic Journeys* (Minneapolis, MN: Fortress Press, 2005).
11. Sarah McFarland Taylor, *Ecopiety: Green Media and the Dilemma of Environmental Virtue* (New York: New York University Press, 2019).
12. Mary-Jane Rubenstein, *Strange Wonder: The Closure of Metaphysics and the Opening of Awe* (New York: Columbia University Press, 2008), 10–11.
13. Lorraine Daston and Katherine Park, *Wonders and the Order of Nature, 1150–1750* (New York: Zone Books, 1998).
14. This is the argument in Whitney Bauman, *Theology, Creation and Environmental Ethics: From* Creatio ex Nihilo *to* Terra Nullius (New York: Routledge, 2009).
15. Virginia Burrus, *Ancient Christian Ecopoetics: Cosmologies, Saints, Things* (Philadelphia: University of Pennsylvania Press, 2019); Kate Rigby, *Meditations on Creation in an Era of Extinction* (Maryknoll, MD: Orbis Books, 2023).
16. See, e.g., Mary-Jane Rubenstein, *Pantheologies: God, World, Monsters* (New York: Columbia University Press, 2018); and Catherine Keller, *Cloud of the Impossible: Negative Theology and Planetary Entanglement* (New York: Columbia University Press, 2014).
17. Keller, *Cloud of the Impossible*, 121.
18. Louise Economides, *The Ecology of Wonder in Romantic and Postmodern Literature* (New York: Palgrave Macmillan, 2016), 1.
19. Marcus Boon, Eric Cazdyn, and Timothy Morton, *Nothing: Three Inquiries in Buddhism* (Chicago: University of Chicago Press, 2015).
20. See, e.g., Brian Greene, *The Fabric of the Cosmos: Space, Time and the Texture of Reality* (New York: Vintage Books, 2004).
21. Keller, *Cloud of the Impossible*, 20.
22. Karen Barad, *Meeting the Universe Halfway: Quantum Physics and the Entanglement of Matter and Meaning* (Durham, NC: Duke University Press, 2007).

23. Fuyuki Kurasawa, "A Cosmopolitanism from Below: Alternative Globalization and the Creation of a Solidarity Without Bounds," *European Journal of Sociology*, 45, no. 2 (2004): 233–255. Bron Taylor, *Dark Green Religion: Nature Spirituality and the Planetary Future* (Berkeley: University of California Press, 2010).
24. Bruno Latour, *Critical Zones: The Science and Politics of Landing on Earth* (Cambridge, MA: MIT Press, 2020). Vandana Shiva, *Earth Democracy: Justice, Sustainability, and Peace* (Berkeley, CA: North Atlantic Books, 2005).
25. Bruno Latour wonders about the breakdown of stories and outlines the problems associated with loss of common stories, in *Down to Earth: Politics in the New Climatic Regime* (Cambridge: Polity Press, 2018).
26. Matthias Fritsch, Philippe Lynes, and David Woods eds., *Eco-Deconstruction: Derrida and Environmental Philosophy* (New York: Fordham University Press, 2018).
27. Whitney Bauman, "Critical Planetary Romanticism: Ecology, Evolution, and Erotic Thinking," *Journal for the Study of Religion, Nature, and Culture* 15, no. 1 (2021): 33–52.
28. Latour, *Politics of Nature*.
29. Bruno Latour, *Facing Gaia: Eight Lectures on the New Climatic Regime* (Cambridge: Polity Press, 2017), 59.
30. I draw from Eduardo Viveiros de Castro in thinking through multinaturalism, but not at the expense of making culture/subject universal. See *The Relative Native: Essays on Indigenous Conceptual Worlds* (Chicago: Hau Books, 2015), 191ff.
31. On Haraway's naturecultures, see *Simians, Cyborgs and Women: The Reinvention of Nature* (New York: Routledge, 1991), 149–182. On Kaufman's biohistory, see his *In Face of Mystery: A Constructive Theology* (Cambridge, MA: Harvard University Press, 1993).
32. Gayatri Spivak, *Death of a Discipline* (New York: Columbia University Press, 2003).
33. See, e.g., Kocku von Stuckrad, *A Cultural History of the Soul: Europe and North America from 1870 to the Present* (New York: Columbia University Press, 2022); and Kate Rigby, *Reclaiming Romanticism: Towards an Ecopoetics of Decolonization* (New York: Bloomsbury, 2021).
34. Johan Wolfgang Goethe, *The Metamorphosis of Plants* (1863; Junction City, OR: Bio-Dynamic Farming and Gardening Association, 2004);

Erasmus Darwin, *The Botanic Garden: Containing the Loves of the Plants. A Poem with Philosophical Notes* (Dublin: J. Moore, 1790).

35. I've written about this elsewhere. see: Whitney Bauman, "Wonder and Ernst Haeckel's Aesthetics of Nature," in *Arts, Religion, and the Environment*, ed. Sigurd Bergmann and Forrest Clingerman (Leiden, the Netherlands: Brill 2018), 61–83.
36. Gustav Theodor Fechner, *Nanna oder über das Seelenleben Der Pflanzen* (Leipzig, Germany: Edition Mabila, 2013); Wilhelm Bölsche, *Love Life in Nature*, 2 vols (New York: Albert and Charles Boni, 1926).
37. Dalia Nassar argues, along similar lines that I hope to argue, that the way in which Romanticism and empiricism have been juxtaposed in the history of science and philosophy of science is not accurate or helpful, but rather that a "romantic empiricism" exists and has a long history. See Dalia Nassar, *Romantic Empiricism: Nature, Art, and Ecology from Herder to Humboldt* (New York: Oxford University Press, 2022).
38. Peter Coates, *American Perceptions of Immigrant and Invasive Species: Strangers on the Land* (Berkeley: University of California Press, 2006).
39. Whitney Bauman, *Religion and Ecology: Developing a Planetary Ethic* (New York: Columbia University Press, 2014).
40. The practice of "creaturely eco-poetics" that Kate Rigby articulates is helpful here. It is a poetics that "highlights human entanglements, at once material and moral, with other living beings. These entanglements entail shared, if unevenly distributed, vulnerabilities as well as shared, if variegated, communicative capacities. They harbour the ever-present risk of conflict and harm but also opportunities to co-create emergent multi-species worlds no longer constrained by the colonizing logic of human-nonhuman hyperseparation, and hence conducive to more felicitous forms of coexistence and 'sympoiesis' (a term coined by Friedrich Schlegel) in our own perilous times"; Rigby, *Reclaiming Romanticism*, 38.
41. Mary-Jane Rubenstein, *Astrotopia: The Dangerous Religion of the Corporate Space Race* (Chicago: University of Chicago Press, 2022).
42. Mary Midgley, *Science and Poetry* (New York: Routledge, 2001).
43. Rob Nixon, *Slow Violence and the Environmentalism of the Poor* (Cambridge, MA: Harvard University Press, 2013).
44. Lisa Sideris and Kathleen Dean Moore, *Rachel Carson: Legacy and Challenge* (Albany: State University of New York Press, 2008), 6.

45. She talks about this in her book *The Sea Around Us*, which was made into a documentary film by the same name that won the Oscar for Best Documentary in 1963. As one reviewer notes, "The film . . . closes at the polar regions, pointing out that the mountains of ice are slowly melting as the climate heats up and that if they should melt entirely the oceans would rise to a height that would inundate all the coast lines. After which could easily come—as the closing title reads—'the end?'"; Irwin Allen, "Review of *The Sea Around Us*," *Hollywood Reporter*, January 15, 1953.
46. Evelyn Fox Keller, *A Feeling for the Organism: The Life and Work of Barbara McKlintock* (New York: W. H. Freeman, 1983).
47. Sideris and Moore, *Rachel Carson*, 8.
48. Rachel Carson, *Under the Sea Wind* (New York: Penguin, 1941), 109.
49. Catherine Keller, *Face of the Deep: A Theology of Becoming* (New York: Routledge, 2003).
50. Carson, *Under the Sea Wind*, 119.
51. Val Plumwood, *Environmental Culture: The Ecological Crisis of Reason* (New York: Routledge, 2001).
52. Ursula Goodenough, *Sacred Depths of Nature: How Life Has Emerged and Evolved*, 2nd rev. ed. (New York: Oxford University Press, 2023).
53. Ernst Bloch, *The Principle of Hope*, vol. 1 (Boston: MIT Press, 1986.).
54. Hannah Arendt, *The Human Condition* (Chicago: University of Chicago Press, 1958).

AFTERWORD: CRITICALLY NAVIGATING THE PLANETARY SEAS

Part of this afterword was developed in Whitney Bauman, "Florida Is Burning: A Call for Connective Knowledge Systems," blog for *Counterpoint: Navigating Knowledge*, February 16, 2023, www.counterpointknowledge.org/florida-is-burning-a-call-for-connective-knowledge-systems/.

1. Vandana Shiva, *Earth Democracy: Justice, Sustainability and Peace* (Berkeley, CA: North Atlantic Books, 2005); Bron Taylor talks about terrapolitanism in his *Dark Green Religion: Nature Spirituality and the Planetary Future* (Berkeley: University of California Press, 2010).

2. Hannah Arendt, *The Human Condition* (Chicago: University of Chicago Press, 1958).
3. The "Mint Julep" model of education refers to US Reconstruction after the Civil War. Textbook makers would make two versions of history books: one which dealt with slavery (for the North) and the other which largely did not (for the Southern states). The latter later came to be known as the "Mint Julep" editions. See Chara Haeussler Bohan, Lauren Yarnell Bradshaw, and Wade Hampton Morris Jr., "The Mint Julep Consensus: An Analysis of Late 19th Century Southern and Northern Textbooks and Their Impact on the History Curriculum," *Journal of Social Studies Research* 44, no. 1 (January 2020): 139–149.
4. Matthews, *For the Love of Matter: A Contemporary Panpsychism* (Albany: State University of New York Press, 2003), 44.
5. Nelle Morton, *The Journey Is Home* (Boston: Beacon, 1985).

BIBLIOGRAPHY

Advancing the Participatory Sciences. Accessed January 15, 2026. https://participatorysciences.org/.

Aescht, Erna, Gerhard Aubrecht, Erika Krauße, and Franz Speta, eds. *Welträtsel und Lebenswunder: Ernst Haeckel—Werk, Wirkung, und Folgen*. Linz, Germany: OO Landesmuseums, 1998.

Allen, Irwin. Review of *The Sea Around Us*. *Hollywood Reporter*, January 15, 1953.

Althaus-Reid, Marcella. *Indecent Theology: Theological Perversions in Sex, Gender, and Politics*. New York: Routledge, 2007.

Anderson, Benedict. *Imagined Communities: Reflections on the Origin and Spread of Nationalism*. New York: Verso, 1983.

Arendt, Hannah. *The Human Condition*. Chicago: University of Chicago Press, 1958.

——. *The Life of the Mind*. New York: Harcourt, 1978.

——. *Sokrates. Apologie der Pluralität*. Berlin: Matthew and Seitz, 2016.

Bach, Thomas, Christina Brandt, and Florence Vienne, eds. *Exploring Ernst Haeckel's Work in the 21st Century*. University of Pittsburgh Press, forthcoming

Bährmann, Rudolf. 2006. "Haeckel's ökologie-Begriff: Inhalt, Deutung, Bedeutung." In *Darwin, Haeckel und die Folgen: Monismus in Vergangenheit und Gegenwart*, ed. Arnher E. Lenz und Volker Mueller, 93–126. Neustadt am Rübenberge, Germany: Angelika Lenz Verlag, 2006.

Barad, Karen. *Meeting the Universe Halfway: Quantum Physics and the Entanglement of Matter and Meaning*. Durham, NC: Duke University Press, 2007.

Bastian, Michelle. "Encountering Leatherbacks in Multispecies Knots of Time." In *Extinction Studies: Stories of Time, Death and Generations*, ed. T. van Dooren, D. Bird Rose, and M. Chrulew, 149–186. New York: Columbia University Press, 2017.

Bauman, Whitney. "Climate Weirding and Queering Nature: Getting Beyond the Anthropocene." *Religions* 6 (2015): 742–754.

——. "Critical Planetary Romanticism: Ecology, Evolution, and Erotic Thinking." *Journal for the Study of Religion, Nature, and Culture* 15, no. 1 (2021): 33–52.

——. "Developing a Critical Planetary Romanticism: Re-Attuning to the Earth." In *Religion, Materialism, and Ecology*, ed. Sigurd Bergmann, Kate Rigby, and Peter Manley Scott, 13–28. New York: Routledge, 2023.

——. "Developing a Critical Romanticism" (blog). *Counterpoint: Navigating Knowledge*, August 14, 2019. www.counterpointknowledge.org/developing-a-critical-romanticism/.

——. "Ernst Haeckel's Creation: Developing a Non-Reductive Religious Naturalism." In *The Routledge Handbook of Religious Naturalism*, ed. Jerome Stone and Donald Crosby, 33–42. New York: Routledge, 2018.

——. "Florida Is Burning: A Call for Connective Knowledge Systems" (blog). *Counterpoint: Navigating Knowledge*, February 16, 2023. www.counterpointknowledge.org/florida-is-burning-a-call-for-connective-knowledge-systems/.

——. "Prismatic Identities in a Planetary Context." In *Ecological Solidarities: Mobilizing Faith and Justice for an Entangled World*, ed. Dhawn Martin, Elaine Padilla, and Krista Hughes, 186–200. University Park: Penn State University Press, 2019.

——. *Religion and Ecology: Developing a Planetary Ethic*. New York: Columbia University Press, 2014.

——. *Theology, Creation and Environmental Ethics: From* Creatio ex Nihilo *to* Terra Nullius. New York: Routledge, 2009.

——. "Queering Authority in Science and Religion." In *Unsettling Science and Religion: Contributions and Questions from Queer Theory*, ed. Lisa Stenmark and Whitney Bauman, 89–106. Lanham, MD: Lexington Books, 2018.

——. "Wonder and Ernst Haeckel's Aesthetics of Nature." In *Arts, Religion and the Environment: Exploring Nature's Texture*, ed. Sigurd Bergmann and Forrest Clingerman, 61–83. Leiden, the Netherlands: Brill, 2018.

Bauman, Whitney, and Kevin O'Brien. *Environmental Ethics and Uncertainty: Wrestling with Wicked Problems.* New York: Routledge, 2019.

Bauman, Zygmunt. *Liquid Love: On the Frailty of Human Bonds.* Cambridge: Polity Press, 2003.

Bennett, Jane. *Vibrant Matter: A Political Ecology of Things.* Durham, NC: Duke University Press, 2010.

Berry, Thomas. *The Great Work: Our Way into the Future.* New York: Random House, 1999.

Bhabha, Homi K. *The Location of Culture.* New York: Routledge, 1984.

Bloch, Ernst. *The Principle of Hope*, vol. 1. Trans. Neville Plaice and Stephen Plaice. Boston: MIT Press, 1986.

Bohan, Chara Haeussler, Lauren Yarnell Bradshaw, and Wade Hampton Morris Jr. "The Mint Julep Consensus: An Analysis of Late 19th Century Southern and Northern Textbooks and their Impact on the History Curriculum." *Journal of Social Studies Research* 44, no. 1 (January 2020): 139–149.

Bölsche, Wilhelm. *Love Life in Nature: The Story of the Evolution of Love.* 2 vols. New York: Albert and Charles Boni, 1926.

——. *Von Wundern und Tieren: Neue Naturwissenschaftliche Plaudereien.* Berlin: Deutsche Verlags-Anstalt, 1920.

Boon, Marcus, Eric Cazdyn, and Timothy Morton. *Nothing: Three Inquiries in Buddhism.* Chicago: University of Chicago Press, 2015.

Bray, Karen. *Grave Attending: A Political Theology for the Unredeemed.* New York: Fordham University Press, 2020.

Bray, Karen, Heather Eaton, and Whitney Bauman. *Earthly Things: Immanence, New Materialisms, and Planetary Thinking.* New York: Fordham University Press, 2023.

Breidbach, Olaf. "Brief Instructions to Viewing Haeckel's Pictures." In *Art Forms in Nature: The Prints of Ernst Haeckel*, by Ernst Haeckel, 9–18. Munich: Prestel-Verlag, 2015.

——. *Visions of Nature: The Art and Science of Ernst Haeckel.* Munich: Prestel, 2006.

Brennan, Teresa 2003. *Globalization and Its Terrors.* New York: Routledge.

Burkeman, Oliver. *Four Thousand Weeks: Time Management for Mortals.* New York: MacMillan, 2021.

Burrus, Virginia. *Ancient Christian Ecopoetics: Cosmologies, Saints, Things.* Philadelphia: University of Pennsylvania Press, 2019.

Butler, Judith. *Bodies That Matter: On the Discursive Limits of Sex*. New York: Routledge, 1993.

——. *Gender Trouble: Feminism and the Subversion of Identity*. New York: Routledge, 1990.

Callicott, J. Baird. *Thinking Like a Planet: The Land Ethic and the Earth Ethic*. New York: Oxford University Press, 2013.

Carey, Brycchan. "Clerical Naturalists in the Age of Enlightenment." www.brycchancarey.com. Accessed September 10, 2025.

——. *The Unnatural Slave Trade: Slavery, Abolition and Environmental Writing, 1650–1807*. New Haven, CT: Yale University Press, 2024.

Carson, Rachel. *Silent Spring*. New York: Mariner, 1962.

——. *Under the Sea Wind*. New York: Penguin, 1941.

Chen, Mel. *Intoxicated: Race, Disability, and Chemical Intimacy Across Empire*. Durham, NC: Duke University Press, 2023.

Coates, Peter. *American Perceptions of Immigrant and Invasive Species: Strangers on the Land*. Berkley: University of California Press, 2006.

Connolly, William. *Climate Machines, Fascist Drives, and Truth*. Durham, NC: Duke University Press, 2019.

Crockett, Clayton, and Jeffrey Robbins. *Religion, Politics, and the Earth: The New Materialism*. New York: Palgrave, 2012.

Darwin, Erasmus. *The Botanic Garden. Containing the Loves of Plants. A Poem. With Philosophical Notes*. Dublin: J. Moore, 1790.

Daston, Lorraine, and Katherine Park. *Wonders and the Order of Nature, 1150–1750*. New York: Zone Books, 1998.

De Castro, Eduardo Viveiros. *The Relative Native: Essays on Indigenous Conceptual Worlds*. Chicago: Hau Books, 2015.

Deacon, Terrence. *Incomplete Nature: How Mind Emerged from Matter*. New York: Norton, 2012.

Deleuze, Gilles, and Felix Guattari. *A Thousand Plateaus: Capitalism and Schizophrenia*. Trans. Brian Massumi. Minneapolis: University of Minnesota Press, 1987.

Derrida, Jacques. *Of Grammatology*. Trans. Gayatri Spivak. Baltimore: John Hopkins University Press, 1974.

——. *The Politics of Friendship*. Trans. George Collins. New York: Verso, 1994.

——. *The Specters of Marx: The State of the Debt, the Work of Mourning*. Trans. Peggy Kamuf. New York: Routledge, 1994.

Di Gregoria, Mario A. *From Here to Eternity: Ernst Haeckel and Scientific Faith*. Göttingen, Germany: Vandenhoeck and Ruprecht, 2005.

Duvernoy, Russell. *Affect and Attention After Deleuze and Whitehead: Ecological Attunement*. Edinburgh: Edinburgh University Press, 2021.

Economides, Louise. *The Ecology of Wonder in Romantic and Postmodern Literature*. New York: Palgrave Macmillan, 2016.

Einboden, Jeffrey. *Islam and Romanticism: Muslim Currents from Goethe to Emerson*. London: One World, 2014.

Ellis, Erika. "The Reception of Ernst Haeckel in America: Monism and The Prophet of Evolution." Undergraduate thesis, University of Washington, Seattle, 1998.

Elsner, Norber. *Ungelöste Welträtsel: Frida von Uslar-Gleichen und Ernst Haeckel Briefe und Tagebücher 1898–1903, Band 3*. Göttingen, Germany: Wallstein Verlag, 1996.

Fechner, Gustav Theodor. *Nanna oder über das Seelenleben der pflanzen*. Germany: Edition Mabila, 2013.

Feuerbach, Ludwig. *The Essence of Christianity*. London: Paul, Trench and Trübner, 1890.

Foucault, Michel. *Discipline and Punish: The Birth of the Prison*. Trans. Alan Sheridan. New York: Random House, 1977.

——. *The Order of Things: An Archeology of the Human Sciences*. New York: Random House, 1970.

——. *Power/Knowledge: Selected Interviews and Other Writings, 1972–1977*. Trans. Colin Gordon. New York: Random House, 1988.

Fritsch, Matthias, Philippe Lynes, and David Woods, eds. *Eco-Deconstruction: Derrida and Environmental Philosophy*. New York: Fordham University Press, 2018.

Gasman, Daniel, ed. *The Scientific Origins of National Socialism*. New York: Routledge, 2017.

Gebara, Ivone. *Longing for Running Water: Ecofeminism and Liberation*. Trans. David Molineaux. Minneapolis, MN: Fortress Press, 1999.

Gilboff, Sander. *H. G. Bronn, Ernst Haeckel and the Origins of German Darwinism: A Study in Translation and Transformation*. Cambridge, MA: MIT Press, 2008.

Goethe, Johan Wolfgang. *The Metamorphosis of Plants*. Trans. Gordon Miller. Junction City, OR: Bio-Dynamic Farming and Gardening Association, 2004. First published in 1863.

——. *Theory of Colors*. London: Trans. Charles Lock Eastlake. John Murray Press, 1840.

Goodenough, Ursula. *The Sacred Depths of Nature: How Life Has Emerged and Evolved*. 2nd rev. ed. New York: Oxford University Press, 2023.

Goodenough, Ursula, and Terrence Deacon. "The Sacred Emergence of Nature." In *The Oxford Handbook of Science and Religion*, ed. Phillip Clayton, 853–871. New York: Oxford University Press, 2008.

Gould, Stephen J. *Ontogeny and Phylogeny*. Cambridge, MA: Harvard University Press, 1977.

Greene, Brian. *The Fabric of the Cosmos: Space, Time and the Texture of Reality*. New York: Vintage Books, 2004.

Grigg, Russel. "Ernst Haeckel, Evangelist for Evolution and Apostle of Deceit." Creation Ministries International, January 27, 2006, updated February 4, 2015. http://creation.com/ernst-haeckel-evangelist-for-evolution-and-apostle-of-deceit. First published in *Creation* 18, no. 2 (1996): 33–36.

Haeckel, Ernst. *Art Forms in Nature: The Prints of Ernst Haeckel*. Munich: Prestel-Verlag, 2015.

——. *Ausgewählte Briefwechsel. Historisch-kritische Ausgabe. Band 1: Familien Korrespondenz Februar 1839–April 1845*. Ed. Roman Göbel, Gerhard Müller, Claudia Taszus, Thomas Bach, Jens Pahnke and Kathrin Polenz. Stuttgart, Germany: Franz Steiner, 2017.

——. *Gott-Natur (Theophysis), mit Kommenter Nachdruck von Olaf Breidbach und Uwe Hoßfeld*. Stuttgart, Germany: Franz Steiner Verlag, 2008.

——. *History of Creation: Or the Development of the Earth and Its Inhabitants by the Action of Natural Causes*, vol. 1. Trans. E. Ray Lankester. New York: D. Appleton, 1876.

——. *Kristallseelen: Studien über das anorganische Leben*. Leipzig, Germany: Alfred Kroner Verlag, 1917.

——. *The Last Link: Our Present Knowledge of the Descent of Man*. London: Adam and Charles Black, 1898.

——. *Monism as Connecting Religion and Science: The Confession of Faith of a Man of Science*. Trans. J. Gilchrist. London: Adam and Charles Black, 1894.

——. *The Riddle of the Universe at the Close of the Nineteenth Century*. Trans. Joseph McCabe. New York: Harper, 1900.

——. *The Wonders of Life: A Popular Study of Biological Philosophy*. Trans. Joseph McCabe. New York: Harper, 1904.

Halberstam, Jack. *The Queer Art of Failure*. Durham, NC: Duke University Press, 2011.

——. *Wild Things: The Disorder of Desire*. Durham, NC: Duke University Press, 2020.

Haraway, Donna. *Simians, Cyborgs and Women: The Reinvention of Nature*. New York: Routledge, 1991.

——. "Situated Knowledges: The Science Question in Feminism and the Privilege of Partial Perspective." *Feminist Studies* 14, no. 3 (Autumn 1988): 575–599.

——. *Staying With the Trouble: Making Kin in the Chthulucene*. Durham, NC: Duke University Press, 2016.

——. *When Species Meet (Posthumanities)*. Minneapolis: University of Minnesota Press, 2008.

Harding, Sandra. *Objectivity and Diversity: Another Logic of Scientific Research*. Chicago: University of Chicago Press, 2015.

——, ed. *The Postcolonial Science and Technology Studies Reader*. Durham, NC: Duke University Press, 2011.

Harman, Graham. *Object-Oriented Ontology: A New Theory of Everything*. London: Pelican Books, 2018.

Harney, Stefano, and Fred Moten. *The Undercommons: Fugitive Planning and Black Study*. New York: Autonomedia, 2013.

Harrison, Peter. *The Territories of Science and Religion*. Chicago: University of Chicago Press, 2015.

Heidegger, Martin. *The Question Concerning Technology and Other Essays*. New York: Harper and Row, 1977.

Hirschfeld, Magnus. "Ernst Haeckel und die Sexualwissenschaft." In *Was wir Ernst Haeckel verdanken: Ein Buch der Verehrung und Dankbarkeit, 2 Band*, ed. Heinrich Schmidt, 278–282. Leipzig, Germany: Verlag Unesma, 1914.

Hogue, Michael. *The Tangled Bank: Toward an Ecotheological Ethics of Responsible Participation*. Eugene, OR: Wipf and Stock, 2008.

Hohlwein, H. "Keplerbund." In *Die Religion in Geschichte und Gegenwart*, Band 3. Tübingen, Germany: JCB Mohr, 1959.

Holt, Robert Niles. "The Social and Political Ideas of the German Monist Movement, 1871–1914." PhD dissertation, Yale University, 1967.

Horst, W. J. Rittel, and Melvin M. Webber. "Dilemmas in a General Theory of Planning." *Policy Sciences* 4, no. 2 (June 1973.): 155–169.

Horkheimer, Max, and Theodor Adorno. *Dialectic of Enlightenment*. Stanford, CA: Stanford University Press, 2002.

Jacobsen, Eric Paul. *From Cosmology to Ecology: The Monist Worldview in Germany from 1770 to 1930*. Bern: Peter Lang, 2005.

Kaiser, David. *How the Hippies Saved Physics: Science, Counterculture, and the Quantum Revival*. New York: Norton, 2011.

Kaufmann, Gordon. *In Face of Mystery: A Constructive Theology*. Cambridge, MA: Harvard University Press, 1993.

Keller, Catherine. *Cloud of the Impossible: Negative Theology and Planetary Entanglement*. New York: Columbia University Press, 2014.

——. *Face of the Deep: A Theology of Becoming*. New York: Routledge, 2003.

——. *God and Power: Counterapocalyptic Journeys*. Minneapolis, MN: Fortress, 2005.

——. *A Political Theology for the Earth*. New York: Columbia University Press, 2018.

Keller, Catherine and Laurel Kearns. "Introduction: Grounding Theory-Earth in Religion and Philosophy." In *EcoSpirit: Religions and Philosophies for the Earth*, ed. Laurel Kearns and Catherine Keller, 1–20. New York: Fordham University Press, 2007.

Keller, Catherine, and Laurel Schneider, eds. *Polydoxy: Theology of Multiplicity and Relation*. New York: Routledge, 2011.

Keller, Evelyn Fox. *A Feeling for the Organism: The Life and Work of Barbara McCllintock*. New York: W. H. Freeman, 1983.

Kirby, Benjamin. "Religious Infrastructure: Designations, Transformations, Entanglements." *Religion, State and Society* 52, nos. 2–3 (2024): 96–113.

Ko, Aph. *Racism as Zoological Witchcraft: A Guide to Getting Out*. Brooklyn, NY: Lantern Books, 2019.

Kockerbeck, Christoph, ed. *Carl Vogt, Jacob Moleschott, Ludwig Büchner, Ernst Haeckel: Briefwechsel*. Marburg, Germany: Basilisken-Presse, 1999.

Kohn, Eduardo. *How Forests Think: Toward an Anthropology Beyond the Human*. Berkeley: University of California Press, 2013.

Kovach. Margaret. *Indigenous Methodologies: Characteristics, Conversations and Contexts*. 2nd ed. Toronto: University of Toronto Press, 2021.

Krauße, Erika. *Ernst Haeckel: Biographien Hervorragender Naturwissenschaftler, Techniker und Mediziner*. Leipzig, Germany: Teubner Verlagsgesellschaft, 1987.

Kurasawa, Fuyuki. "A Cosmopolitanism from Below: Alternative Globalization and the Creation of a Solidarity Without Bounds." *European Journal of Sociology*. 45, no. 2 (2004): 233–255.

Latour, Bruno. *After Lockdown: A Metamorphosis*. Trans. Julie Ross. Cambridge: Polity Press, 2021.

——. *Critical Zones: The Science and Politics of Landing on Earth*. Cambridge, MA: MIT Press, 2020.

——. *Down to Earth: Politics in the New Climactic Regime*. Cambridge: Polity Press, 2018.

——. *Facing Gaia: Eight Lectures on the New Climatic Regime*. Trans. Catherine Porter. Cambridge: Polity Press, 2017.

——. *If We Lose the Earth, We Lose Our Souls*. Trans. Catherine Porter. Cambridge: Polity Press, 2024.

——. *The Politics of Nature: How to Bring the Sciences into Democracy*. Translated by Catherine Porter. Cambridge, MA: Harvard University Press, 2004.

——. "Thou Shalt Not Freeze-Frame, or How Not to Misunderstand the Science and Religion Debate." In *Science, Religion and the Human Experience*, ed. James Proctor, 27–48. New York: Oxford University Press, 2005.

Lehnert, Erik. "Tiefes Gemüt, klarer Verstand und tapfere Kulturarbeit: Bruno Wille Kaiserreich." In *Darwin Haeckel und die Folgen: Monisumus in Vergangenheit und Gegenwart*, 247–274. Neustadt am Rübenberge, Germany: Angelika Lenz Verlag, 2006.

Manning, Erin. "Another Regard." In *The Lure of Whitehead*, ed. Nicholas Gaskill and A. J. Nocek, 312–331. Minneapolis: University of Minnesota Press, 2014.

Marks, Jonathan. "Why Were the First Anthropologists Creationists?" *Evolutionary Anthropology* 19 (2010): 222–226.

Matthews, Freya. *For the Love of Matter: A Contemporary Panpsychism*. Albany: State University of New York Press, 2003.

McNeill, J. R. *The Great Acceleration: An Environmental History of the Anthropocene Since 1945*. Cambridge, MA: Harvard University Press, 2014.

Merchant, Carolyn. *Autonomous Nature: Problems of Prediction and Control from Ancient Times to the Scientific Revolution*. New York: Routledge, 2015.

——. *The Death of Nature: Women, Ecology and the Scientific Revolution*. New York: HarperCollins, 1980.

——. *Reinventing Eden: The Fate of Nature in Western Culture*. New York: Routledge, 2003.

Midgley, Mary. *Science and Poetry*. New York: Routledge, 2002.

Mignolo, Water. *The Dark Side of the Renaissance: Literacy, Territoriality, and Colonization*. Ann Arbor: University of Michigan Press, 2003.

——. *The Darker Side of Western Modernity: Global Futures, Decolonial Options*. Durham, NC: Duke University Press, 2011.

Miller, James. *China's Green Religion: Daoism and the Quest for a Sustainable Future*. New York: Columbia University Press, 2017.

Millsapps, Jan. "The Arduous Journey of Staying Home" (blog). *Counterpoint, Navigating Knowledge*, March 26, 2025. www.counterpointknowledge.org/the-arduous-journey-of-staying-home/.

Minister, Kevin. "Interreligious Approaches to Sustainability Without a Future: Two New Materialist Proposals for Religion and Ecology." In *Earthly Things: Immanence, New Materialisms and Planetary Thinking*, ed. Karen Bray, Heather Eaton, and Whitney Bauman, 136–147. New York: Fordham University Press, 2023.

Morton, Marsha. "From Monera to Man: Ernst Haeckel, Darwinismus, and Nineteenth Century German Art" In *The Art of Evolution: Darwin, Darwinismus and Visual Culture*, ed. Barbara Larson and Fae Brauer, 59–91. Hanover, NH: Dartmouth College Press, 2009.

Morton, Nelle. *The Journey Is Home*. Boston: Beacon, 1985.

Morton, Timothy. *Dark Ecology: For a Logic of Future Co-Existence*. New York: Columbia University Press, 2016.

——. *Hyperobjects: Philosophy and Ecology After the End of the World*. Minneapolis: University of Minnesota Press, 2013.

Nancy, Jean-Luc. *Being Singular Plural*. Trans. by Robert D. Richardson and Anne E. O'Byrne. Stanford, CA: Stanford University Press, 2000.

Nassar, Dalia. *The Romantic Absolute: Being and Knowing in Early German Romantic Philosophy, 1795–1804*. Chicago: University of Chicago Press, 2014.

——. *Romantic Empiricism: Nature, Art and Ecology from Herder to Humboldt*. New York: Oxford University Press, 2022.

Nichols, Amanda, and Whitney Bauman. "The Watery Depths of American Environmentalism: Marjory Stoneman Douglas, Rachel Carson, and Sylvia Earle." *Journal for the Study of Religion, Nature and Culture* 16, no. 3 (2022): 294–425.

Nietzsche, Friedrich. *The Gay Science*. Trans. Walter Kaufmann. New York: Vintage, 1974.

——. *Thus Spake Zarathustra*. Trans. Thomas Common. New York: Macmillan, 1911.

Nixon, Rob. *Slow Violence and the Environmentalism of the Poor*. Cambridge, MA: Harvard University Press, 2013.

Nötlich, Rosemarie. *Ernst Haeckel-Wilhelm Bölche: Breifwechsel 1887–1919*. Berlin: Verlag für Wissenschaft und Bildung, 2022.

Peterson, Anna. *Being Human: Ethics, Environment and Our Place in the World*. Berkeley: University of California Press, 2001.

Plumwood, Val. *Environmental Culture: The Ecological Crisis of Reason*. New York: Routledge, 2002.

Princen, Thomas, Jack Manno, and Pamela Martin, eds. *Ending the Fossil Fuel Era*. Cambridge, MA: MIT Press, 2015.

Richards, Robert. "Ernst Haeckel and the Struggles over Evolution and Religion." *Annals of the History of Philosophy and Biology* 10 (2005): 89–115.

——. "The German Reception of Darwin's Theory, 1860–1945." In *The Cambridge Encyclopedia of Darwin and Evolutionary Thought*, ed. 235–242. Cambridge: Cambridge University Press, 2013.

——. *The Romantic Conception of Life: Science and Philosophy in the Age of Goethe*. Chicago: University of Chicago Press, 2002.

——. *The Tragic Sense of Life: Ernst Haeckel and the Struggle over Evolutionary Thought*. Chicago: University of Chicago Press, 2009.

Rigby, Kate. *Meditations on Creation in an Era of Extinction*. Maryknoll, MD: Orbis Press, 2023.

——. "Nature, Language, and Religion: Herder and Beyond." In *Ecological Thought in German Literature and Culture*, ed. G. Duerbeck, U. Stobbe, H. Zapf, and E. Zemanek, 31–41. Lanham, MD: Lexington Books, 2017.

——. *Reclaiming Romanticism: Towards an Ecopoetics of Decolonization*. Oxford: Bloomsbury, 2021.

——. *Topographies of the Sacred: The Poetics of Place in European Romanticism*. Charlottesville: University of Virginia Press, 2004.

Rivera, Mayra. *Poetics of the Flesh*. Durham, NC: Duke University Press, 2015.

Rosa, Hartmut. *Resonance: A Sociology of Our Relationship to the World*. Cambridge: Polity Press, 2019.

Rosenkranz, Karl. *Aesthetics of the Ugly*. London: Bloomsbury, 2015.

Ross, Loretta. *Calling In: How to Start Making Changes with Those You'd Rather Cancel.* New York: Simon and Schuster, 2025.

Rowe, Terra Schwerin. *Of Modern Extraction: Experiments in Critical Petro-Theology.* London: T&T Clark, 2022.

Rubenstein, Mary-Jane. *Astrotopia: The Dangerous Religion of the Corporate Space Race.* Chicago: University of Chicago Press, 2022.

——. *Pantheologies: Gods, Worlds, Monsters.* New York: Columbia University Press, 2018.

——. *Strange Wonder: The Closing of Metaphysics and the Opening of Awe.* New York: Columbia University Press, 2008.

Ruether, Rosemary Radford. *Integrating Ecofeminism, Globalization and World Religions.* Lanham, MD: Roman and Littlefield, 2005.

Rushkoff, Douglas. *Survival of the Richest: Escape Fantasies of the Tech Billionaires.* London: Scribe Publications, 202

Sandmann, Jürgen. *Der Bruch mit der Humanitären Tradition: Die Biologisierung der Ethik bei Ernst Haeckel und anderen Darwinisten seiner Zeit.* Stuttgart, Germany: Gustav Fischer Verlag, 1990.

Saul, Nicholas. "'Vorgänger Darwins' or Nachfolger Goethes? Willhelm Bölsche on Evolutionism in Goethe and Novalis or, Literature and the Two Cultures Then and Now." *Publication of the English Goethe Society*, 84, no. 1 (2015): 76–91.

Schaefer, Donovan. *Religious Affects: Animality, Evolution and Power.* Durham, NC: Duke University Press, 2015.

——. "The Things of Order: Affect, Material Culture, *Dispositif*." *Cultural Critique*, 124 (Summer 2024.): 1–30.

——. *Wild Experiment: Feeling Science and Secularism After Darwin.* Durham, NC: Duke University Press, 2022.

Schelling, Friedrich Wilhelm. J. *Bruno: On the Natural and Divine Principle of Things.* Trans. Michael Vater. Albany: State University of New York Press Press, 2000.

——. *On the World Soul.* Hamburg, Germany: Perthes, 1798.

Schiebinger, Londa. *Nature's Body: Gender in the Making of Modern Science.* Boston: Beacon Press, 1993.

Schleiermacher, Friedrich. *The Christian Faith.* London: T&T Clark, 2016.

——. *On Religion: Speeches to Its Cultured Despisers.* Christian Classics Ethereal Library. 1799. Accessed January 15, 2026. www.ccel.org/ccel/s/schleiermach/religion/cache/religion.pdf.

Schmidt, Heinrich, ed. *Was Wir Ernst Haeckel Verdanken*. 2 vols. Dresden, Germany: Fachbuchverlag, 2014.

Schmitt. Carl. *The Concept of the Political*. Chicago: University of Chicago Press, 1996.

Schwarz, Hans. "Darwinism Between Kant and Haeckel." *Journal of the American Academy of Religion* 40 (1980): 581–602.

Seidel, Andrea Mantell. *Isadora Duncan in the 21st Century: Capturing the Art and Spirit of the Dancer's Legacy*. Jefferson, NC: McFarland, 2016.

Shiva, Vandana. *Earth Democracy: Justice, Sustainability and Peace*. Berkeley, CA: North Atlantic Books, 2005.

——. *Staying Alive: Women, Ecology and Development*. London: Zed Books, 1988.

Sideris, Lisa. *Consecrating Science: Wonder, Knowledge and the Natural World*. Berkeley: University of California Press, 2017.

Sideris, Lisa, and Kathleen Dean Moore. *Rachel Carson: Legacy and Challenge*. Albany: State University of New York Press Press, 2008.

Spivak, Gayatri. *Death of a Discipline*. New York: Columbia University Press, 2003.

Stengers, Isabelle. "The Cosmopolitical Proposal." In *Making Things Public*, ed. Bruno Latour and Peter Weibel, 994–1003. Cambridge, MA: MIT Press, 2005.

Strauss, David. *Das Leben Jesu: Für das Deutsche Volk Bearbeitet*. Leipzig, Germany: F A Brockhaus, 1864.

Taylor, Bron. *Dark Green Religion: Nature Spirituality and the Planetary Future*. Berkeley: University of California Press, 2010.

Taylor, Charles. *A Secular Age*. Cambridge, MA: Harvard University Press, 2007.

Taylor, Sarah McFarland. *Ecopiety: Green Media and the Dilemma of Environmental Virtue*. New York: New York University Press, 2019.

——. "'F*ck Earth': Unmasking Mars Colonization Marketing, from Planetary Perceived Obsolescence to Apocalyptic 'New Earth' Rhetoric." *Journal of Religion, Media and Culture* 11 (2022): 54–84.

Thompson, Paul. *The Spirit of the Soil: Agriculture and Environmental Ethics*. New York: Routledge, 1995.

Tucker, Mary Evelyn. *Worldly Wonder: Religions Enter Their Ecological Phase*. Peru, IL: Open Court, 2003.

Tweed, Thomas. *Crossing and Dwelling: A Theory of Religion*. Cambridge, MA: Harvard University Press, 2008.

Vollmer, Laura Jean. "The Relationality of Religion and Science: Toward a New Discourse-Analytical Framework." PhD Dissertation, University of Groningen, 2017.

Von Stuckrad, Kocku. *A Cultural History of the Soul: Europe and North America from 1870 to the Present*. New York: Columbia University Press, 2022.

——. *The Scientification of Religion: An Historical Study of Discursive Change, 1800–2000*. Berlin: De Gruyter, 2014.

Weber, Andreas. *Matter and Desire: An Erotic Ecology*. White River Junction, VT: Chelsea Green, 2017.

Weber, Heiko. "Der Monismus als Theorie einer einheitlichen Weltanshauung am Beispiel der Positionen von Ernst Haeckel und August Forel." In *Monismus um 1900: Wissenschaftskultur und Weltanschauung*, ed. Paul Ziche, 81–127. Berlin: Verlag für Wissenschaft und Bildung, 2000.

——. *Monistische und antimonistische Weltanshauung: Eine Auswahlbibliographie*. Berlin: Verlag für Wissenschaft und Bildung, 2000.

Weber, Heiko, and Olaf Breidbach. "Der Deutsche Monistbund: 1906 bis 1933." In *Darwin, Haeckel und die Flogen: Monismus in Vergangenheit und Gegenwart*, ed. A. Lenz and V. Mueller, 157–206. Neustadt am Rübenberg, Germany: Angelika Lenz Verlag, 2006.

Weir, Todd, ed. *Monism: Science, Philosophy, Religion and the History of a Worldview*. New York: Palgrave, 2012.

Wheeler, Wendy. *Expecting the Earth: Life, Culture, Biosemiotics*. London: Lawrence and Wishart, 2016.

White. Carol Wayne. *Black Lives and Sacred Humanity: Toward an African American Religious Naturalism*. New York: Fordham University Press,2016.

Whitehead, Alfred North. *Adventures of Ideas*. New York: Free Press, 1993.

——. *Process and Reality*. New York: Free Press, 1978.

——. *Science and the Modern World*. New York: MacMillan, 1925.

Whitt, Laurelyn. *Science, Colonialism, and Indigenous Peoples: The Cultural Politics of Law and Knowledge*. New York: Cambridge University Press, 2009.

Wohlleben, Peter. *The Hidden Life of Trees: What They Feel, How they Communicate*. Vancouver: Greystone Books, 2016.

Wright, Michelle. *Physics of Blackness: Beyond the Middle Passage Epistemology*. Minneapolis: University of Minnesota Press, 2015.

Wuketits, Franz M. "Mein Lieber Haeckel! Ernst Haeckel, Charles Darwin und der Darwinismus." In *Darwin Haeckel und die Folgen: Monisumus in Vergangenheit und Gegenwart*, 11–32. Neustadt am Rübenberge, Germany. Angelika Lenz Verlag, 2006.

Wulf, Andrea. *The Invention of Nature: Alexander von Humboldt's New World*. New York: Vintage, 2015.

Wynter, Sylvia. *On Being Human as Praxis*. Durham, NC: Duke University Press, 2015.

Zimmerman, Michael. *Heidegger's Confrontation with Modernity: Technology, Politics and Art*. Bloomington: Indian University Press, 1990.

INDEX

Printed and bound by CPI Group (UK) Ltd, Croydon, CR0 4YY

16/09/2026

14963454-0003